Implant Wear in Total Joint Replacement:
Clinical and Biologic Issues,
Material and Design Considerations

Published by the American Academy of Orthopaedic Surgeons

Implant Wear in Total Joint Replacement:
Clinical and Biologic Issues,
Material and Design Considerations

Edited by
Timothy M. Wright, PhD
Senior Scientist, Biomedical Mechanics and Materials
The Hospital for Special Surgery
New York, New York

Stuart B. Goodman, MD, PhD
Professor and Chief of Orthopaedic Surgery
Stanford University
Stanford, California

Symposium
Oakbrook, Illinois
October 2000

Supported by the
National Institute of Arthritis and Musculoskeletal and Skin Diseases

with additional contributions from

American Association of Hip and Knee Surgeons
Biomet, Inc.
The Knee Society
National Institute of Standards and Technology
Orthopaedic Research and Education Foundation
Orthopaedic Research Society
Stryker Howmedica Osteonics

and published by the
American Academy of Orthopaedic Surgeons
6300 North River Road
Rosemont, Illinois 60018

Implant Wear in Total Joint Replacement:
Clinical and Biologic Issues, Material and Design Considerations

The material presented in *Implant Wear in Total Joint Replacement: Clinical and Biologic Issues, Material and Design Considerations* has been made available by the American Academy of Orthopaedic Surgeons for educational purposes only. This material is not intended to present the only, or necessarily best, methods or procedures for the medical situations discussed, but rather is intended to represent an approach, view, statement, or opinion of the author(s) or producer(s), which may be helpful to others who face similar situations.

Some drugs or medical devices demonstrated in Academy courses or described in Academy print or electronic publications have not been cleared by the Food and Drug Administration (FDA) or have been cleared for specific uses only. The FDA has stated that it is the responsibility of the physician to determine the FDA clearance status of each drug or device he or she wishes to use in clinical practice.

Furthermore, any statements about commercial products are solely the opinion(s) of the author(s) and do not represent an Academy endorsement or evaluation of these products. These statements may not be used in advertising or for any commercial purpose.

First Edition
Copyright © 2001 by the American Academy of Orthopaedic Surgeons

ISBN 0-89203-261-8

American Academy of Orthopaedic Surgeons

Contributors

Nicholas Athanasou, MD, PhD*†
Reader in Orthopaedic Pathology
Nuffield Orthopaedic Centre
University of Oxford
Oxford, United Kingdom

Donald L. Bartel, PhD*†
Professor
Mechanical and Aerospace
  Engineering
Cornell University
Ithaca, New York

Thomas W. Bauer, MD, PhD*††
Departments of Pathology and
  Orthopaedic Surgery
The Cleveland Clinic Foundation
Cleveland, Ohio

Joan E. Bechtold, PhD*††
Director of Orthopaedic Research
Midwest Orthopaedic Research
  Foundation
Minneapolis, Minnesota

Matthew J. Beckman, PhD*
Assistant Professor of Biochemistry
  and Orthopaedic Surgery
Medical College of
Virginia/Virginia Commonwealth
  University
Richmond, Virginia

Barbara D. Boyan, PhD*††
Professor of Orthopaedics
Director of Orthopaedic Research
The University of Texas Health
  Science Center at San Antonio
San Antonio, Texas

Thomas D. Brown, PhD*††
Richard C. Johnston Professor of
  Orthopaedic Biomechanics
University of Iowa
Iowa City, Iowa

John J. Callaghan, MD*††
Professor, Department of
  Orthopaedics
University of Iowa
Iowa City, Iowa

Patricia Campbell, PhD*††
Director, Implant Retrieval Lab
Joint Replacement Institute
Orthopaedic Hospital
Los Angeles, California

John C. Clohisy, MD*
Assistant Professor
Department of Orthopaedic Surgery
Washington University
St. Louis, Missouri

David D. Dean, PhD††
Associate Professor of
Orthopaedics
The University of Texas Health
  Science Center at San Antonio
San Antonio, Texas

Avram Allan Edidin, PhD*
Principal Research Scientist
Research Associate Professor
R & D Corporate
Stryker Corporation
Allendale, New Jersey

Gerard A. Engh, MD*††
President
Director, Knee Research
Anderson Orthopaedic Research
Institute
Alexandria, Virginia

Andrew A. Freiberg, MD††
Assistant Professor of Orthopaedic
  Surgery
Massachusetts General Hospital
Boston, Massachusetts

Jorge O. Galante, MD*††
The Grainger Director
The Rush Arthritis and Orthopedics
  Institute
Rush-Presbyterian-St. Luke's
Medical Center
Chicago, Illinois

Victor M. Goldberg, MD*†
Professor and Chairman
Department of Orthopaedics
Case Western Reserve University
University Hospitals of Cleveland
Cleveland, Ohio

Steven R. Goldring, MD*††
Professor of Medicine
Rheumatology
New England Baptist Bone and
  Joint Institute
Harvard Medical School
Boston, Massachusetts

Stuart B. Goodman, MD, PhD*††
Professor and Chief of Orthopaedic
  Surgery
Stanford University
Stanford, California

Ed Greenfield, PhD*
Associate Professor
Department of Orthopaedics
Case Western Reserve University
Cleveland, Ohio

A. Seth Greenwald, DPhil (Oxon)*††
Director of Orthopaedic Research
  and Education
Orthopaedic Research Laboratories
Lutheran Hospital
Cleveland Clinic Health System
Cleveland, Ohio

Nadim Hallab, PhD*
Assistant Professor
Department of Orthopaedic Surgery
Rush-Presbyterian-St. Luke's
Medical Center
Chicago, Illinois

William H. Harris, MD*††
Chief, Adult Reconstructive Service
Massachusetts General Hospital
Boston Massachusetts

Christine S. Heim, BSc††
Engineering Associate
Orthopaedic Research Laboratories
Lutheran Hospital
Cleveland Clinic Health System
Cleveland, Ohio

Stephen M. Hsu, PhD*††
Leader, Surface Properties Group
Materials Science and Engineering
  Lab
National Institute of Standards and
  Technology
Gaithersburg, Maryland

Olga L. Huk, MD*†
Assistant Professor of Surgery
Division of Orthopaedic Surgery
McGill University
Montreal, Canada

Richard Illgen II, MD*
Assistant Professor
University of Wisconsin Medical
  School
Division of Orthopaedic Surgery
Madison, Wisconsin

Joshua J. Jacobs, MD*††
Crown Family Professor
Department of Orthopaedic Surgery
Rush-Presbyterian-St. Luke's
  Medical Center
Chicago, Illinois

Jack E. Lemons, PhD*††
Professor and Director of
  Laboratory Surgical Research
Department of Prosthodontics and
  Biomaterials
Division of Orthopaedic Surgery
University of Alabama at
  Birmingham
Birmingham, Alabama

Stephen Li, PhD*††
President
Medical Device Testing and
  Innovations
Sarasota, Florida

Christoph H. Lohmann, MD*†
Department of Orthopaedics
Georg-August-University
Göttingen, Germany

John M. Martell, MD*
Assistant Professor of Clinical
  Surgery
Section of Orthopaedics and
  Rehabilitative Medicine
The University of Chicago
Chicago, Illinois

Harry A. McKellop, PhD*††
Director, The J. Vernon Luck
  Orthopaedic Research Center
Orthopaedic Hospital
University of California at Los
  Angeles
Los Angeles, California

Donald E. McNulty, PE, MS*
Manager, Tribology
DePuy Orthopaedics, Inc.
Warsaw, Indiana

Mark N. Melkerson, MS*
Deputy Director
Division of General, Restorative,
  and Neurological Devices
Office of Device Evaluation
FDA, Center for Devices and
  Radiological Health
Rockville, Maryland

Gary J. Miller, PhD*
Adjunct Associate Professor
Department of Aerospace,
Engineering, Science, and
  Mechanics
University of Florida
Gainesville, Florida

Orhun Muratoglu, PhD*††
Instructor, Alan Gerry Scholar
Orthopaedic Biomechanics and
  Biomaterials Laboratory
Massachusetts General Hospital
Boston, Massachusetts

Bryan J. Nestor, MD*
Assistant Professor
Department of Orthopaedics
Hospital for Special Surgery
New York, New York

James S. Panagis, MD, MPH*
Director, Orthopaedics Program
National Institute of Arthritis and
  Musculoskeletal and Skin
  Diseases
National Institutes of Health
Bethesda, Maryland

Jack E. Parr, PhD*
Executive Vice-President, Chief
  Scientific Officer
Wright Medical Technology, Inc.
Arlington, Tennessee

Douglas R. Pedersen, PhD†
Research Scientist
Department of Orthopaedic Surgery
University of Iowa
Iowa City, Iowa

Paul D. Postak, BSc††
Engineering Associate
Orthopaedic Research Laboratories
Lutheran Hospital
Cleveland Clinic Health System
Cleveland, Ohio

Claude Rieker, PhD*
Head of Tribology
Sulzer Orthopedics Ltd
Winterthur, Switzerland

Clare M. Rimnac, PhD*††
Associate Professor
Mechanical and Aerospace
  Engineering
Case Western Reserve University
Cleveland, Ohio

Harry E. Rubash, MD*††
Chief, Orthopaedic Surgery
Massachusetts General Hospital
Professor, Harvard Medical School
Boston, Massachusetts

Neil Rushton, MD, FRCS*†
Orthopaedic Research Unit
University of Cambridge
Addenbrooke's Hospital
Cambridge, United Kingdom

Seppo Santavirta, MD, PhD*†
Professor of Orthopaedics and
  Traumatology
University of Helsinki
The Töölö Hospital
Helsinki, Finland

Thomas P. Schmalzried, MD*†
Associate Director
Joint Replacement Institute
Orthopaedic Hospital
Los Angeles, California

David W. Schroeder*
Project Manager, Biomaterials
Biomet, Inc.
Warsaw, Indiana

Zvi Schwartz, DMD, PhD††
Professor, Orthopaedics
The University of Texas Health
  Science Center at San Antonio
San Antonio, Texas

Edward M. Schwarz, PhD*
Assistant Professor, Orthopaedics
University of Rochester
Rochester, New York

Arun S. Shanbhag, PhD*††
Assistant Professor
Department of Orthopaedic Surgery
Massachusetts General Hospital
Harvard Medical School
Boston, Massachusetts

R. Lane Smith, PhD*††
Division of Orthopaedic Surgery
Stanford University School of
  Medicine
Stanford, California

Todd S. Smith*
Director of Materials Research
DePuy Orthopaedics
Warsaw, Indiana

Bernard N. Stulberg, MD*††
Director, Center for Joint
    Reconstruction
Cleveland Orthopaedic and Spine
    Hospital
Lutheran Hospital
Cleveland Clinic Health System
Cleveland, Ohio

Dale Swartz, BS*
Research Department
Zimmer, Inc.
Warsaw, Indiana

John A. Tesk, PhD*
Coordinator Biomaterial Standards
    and Reference Materials
National Institute of Standards and
    Technology
Gaithersburg, Maryland

Peter S. Walker*††
Biomedical Engineering
Cooper Union
New York, New York

Aiguo Wang, PhD*
Director, Tribology
Advanced Technology
Stryker Howmedica Osteonics
Rutherford, New Jersey

Timothy M. Wright, PhD*††
Senior Scientist
Biomedical Mechanics and
    Materials
Hospital for Special Surgery
New York, New York

* Participated in workshop

† Contributed to book

†† Contributed to book and received something of value from a commercial or other party
    related directly or indirectly to subject matter.

Table of Contents

Material and Design Considerations

Preface

Total joint replacement (TJR) for arthritic and degenerative conditions of the large weightbearing joints remains highly successful and cost effective. These operations are being performed in increasing numbers because of the generalized aging of our population and extension of these procedures to younger patients. Despite the success of TJR, failures continue to occur; wear of the implant components is the most common cause. The subsequent revision surgery to replace the worn components is more difficult and costly than the primary operation and has a higher complication rate and a poorer clinical outcome. Safe and effective solutions for prevention, diagnosis, and treatment of wear-related failure require the collaboration of clinicians, bioengineers, material scientists, and biologists in an effort to understand the relevant surgical, biologic, and mechanical problems associated with wear and its attendant bone destruction.

This book summarizes recent collaborative efforts to better understand and address the wear problem. The book is organized around questions that formed the agenda of a workshop sponsored by the National Institutes of Health, the American Academy of Orthopaedic Surgeons, and a number of other organizations, held in Oakbrook, Illinois, October 21-23, 2000. Like the previous NIH/AAOS workshop on wear,[1] questions were initially assigned to individuals, but the responses resulted from collaborative discussion during the workshop and within breakout sessions among all participants. The goals were to set forth current relevant knowledge and future directions for research for a series of clinical, biologic, and engineering aspects of the wear problem. Special emphasis was placed on clinical and basic research and development since the 1995 workshop.

Clinical Issues

Surveillance of TJRs is implemented through periodic clinical and radiographic examinations conducted to detect impending or seminal events that indicate malfunction. Unfortunately, wear of TJRs is generally insidious, with few clinical signs and symptoms until the late stages of failure. Furthermore, wear of implant components is affected by how the patient uses the joint and not merely how long the implant is in service. Thus, risk factors for increased wear and the need for revision surgery include younger, heavier, more active individuals. These patients must be examined more frequently and counseled appropriately. Novel computerized techniques for accurate radiographic measurement of wear hold great promise for early identification of prostheses at risk for failure.

Biologic Issues

The metallic, polymeric, and ceramic particulate byproducts of prosthetic wear are nonbiodegradable. When present in sufficient numbers, wear particles stimulate a foreign body and chronic inflammatory reaction that leads to increased production and release of pro-inflammatory mediators. Proteolytic enzymes, prostanoids, cytokines, chemokines and other substances released into the local environment lead to periprosthetic osteolysis, jeopardizing the long-term stability of the prosthesis. Advances have been made in understanding the cellular and molecular mechanisms associated with loosening and osteolysis; these advances point the way toward strategies for interrupting this biologic cascade.

Material and Design Considerations

Engineering strategies have also emerged to combat the wear problem. Alternative bearing combinations, such as ceramic on ceramic couplings, improved versions of existing materials, such as elevated cross-linked polyethylenes, and improved prosthetic designs hold the promise for significant wear reduction in TJRs. These products are now commercially available based on extensive preclinical in vitro testing and, for alternative bearings, on improvements over earlier, less successful versions of these same bearing couples. However, only long-term clinical observations can determine the value of these improvements.

Timothy M. Wright, PhD
Stuart B. Goodman, MD, PhD

Reference

1. Wright TM, Goodman SB: *Implant Wear: The Future of Total Joint Replacement.* Rosemont, IL, American Academy of Orthopaedic Surgeons, 1996.

Clinical Issues

What is the clinical scope of implant wear in the hip and how has it changed since 1995?

What is the clinical scope of implant wear in the knee and how has it changed since 1995?

What patient-related factors contribute to implant wear?

What surgical-related factors contribute to implant wear?

How should wear-related implant surveillance be carried out and what methods are indicated to diagnose wear-related problems?

What are the systemic consequences of wear debris clinically?

What guidelines/algorithms (both operative and nonoperative) are there for the treatment of osteolysis?

What are the best outcome measures for wear?

What is the outcome of the treatment of osteolysis?

What is the clinical scope of implant wear in the hip and how has it changed since 1995?

Total hip replacement is an effective method of treatment for patients with hip disability.[1] The procedure is capable of providing long-term improvement with excellent control of pain and restoration of function for most patients. An estimated 170,280 total hip replacements were performed in the United States in 1997 (excluding hemiarthroplasties); in 1998, 173,501 were performed. In 1998, 144,133 were primary arthroplasties, a 5% increase over 1994 (Table 1).

The number of revisions was estimated to be 28,794 in 1997 and 29,368 in 1998,[2] representing 17% of total hip replacements and a 7% increase from the revisions reported in 1994. Although 4% of the patients with hip procedures were under age 40, 6% of the revision patients were under 40, pointing to a higher risk of revision for the younger patients. The incidence of revisions in the United States is higher than that reported elsewhere, another potential source of concern.

The indications for total hip replacement have not changed since 1995 and include most conditions causing chronic pain and disability in the hip—osteoarthritis, rheumatoid arthritis, osteonecrosis, posttraumatic arthritis, ankylosing spondylitis, other forms of polyarthritis, some benign and malignant bone tumors, and some types of hip fractures. Active local or systemic infection remains a contraindication for the procedure.

In the past, the best candidates for total hip replacement were thought to be patients over the age of 65 or 70 years; young age was considered a relative contraindication. As more experience has been gained, the indications have been extended progressively to younger patients. Intensive physical activity in a younger individual, however, is generally considered a relative contraindication because the risk of failure increases significantly.

Table 1 Incidence of hip joint arthroplasty in the United States (HCIA 1999 data)

	1998	1997	1994	Increase 1994-1998
Hemi	96,856	93,952	93,439	4%
Primary	144,133	141,486	137,415	5%
Revision	29,368	28,794	27,446	7%
Total	270,357	264,232	258,300	5%

Since the introduction of total hip replacement in the mid 1960s, some significant changes have been introduced in design and materials. For many surgeons, cement is still the preferred method of fixation to the surrounding bone, especially in the femur. Very low long-term failure rates have been reported with traditional Charnley-type implants.[3,4]

Excellent results have also been reported with more modern designs and with the use of modern cementing techniques.[5-7] Over the past 5 years, some concepts related to cemented femoral fixation are being reevaluated. These include issues related to the surface of the femoral stem (such as acrylic pre-coating and surface roughness), the integrity of the cement mantle, and its relation to implant design; all these factors can influence early failure.[8]

Cementless fixation in the femur has replaced the use of acrylic cement in many instances. Excellent intermediate-term results (10 to 15 years) are being reported with second generation design cementless femoral components.[9] One of the major differences between first and second generation implants is that porous coatings are currently applied circumferentially on the surface of the prosthesis,[10] possibly providing a barrier to debris migration. In addition, better design parameters have led to better initial implant stability. Furthermore, these promising results have also been obtained in young patients.

The long-term results with cement fixation in the acetabulum are not as satisfactory as cement fixation with the femur.[11] In addition, 10- and 15-year follow-up studies of cementless acetabulae indicate survival rates superior to those reported with acrylic cement.[12] The use of cementless hemispherical porous-coated implants has largely replaced the use of acrylic cement and is the predominant technique in the United States today.

These different methods of fixation at the femur and the acetabulum may affect both the mechanism and the severity of wear through the production of wear particles. There are multiple potential sources of wear particles: wear of the primary bearing surfaces; wear of one primary surface with another unintended surface, eg, when a femoral component wears through the polyethylene and then articulates against a metal backing; wear when third-body particles become interposed between the two primary bearing surfaces; and wear from the relative motion of a secondary surface against another secondary surface, as can occur at modular interfaces or from implant loosening.[13]

Modularity is an intrinsic feature of every implant design. Major problems have been identified at modular junctions, including fretting and corrosion that contribute to wear, osteolysis, and subsequent implant failure.[14,15] In the past 5 years, design and manufacturing improvements have minimized the potential for these complications at the taper-head junction. The incorporation of additional modularity at the body stem junction in new designs is a source of concern that may be associated with increased wear.

With existing reports it is difficult to stratify revisions of total hip replacement based on the exact cause of failure. Although aseptic loosening remains the most common diagnosis in most reports,[16] the biologic problems associated with wear are the most likely cause of failure, given the advances in fixation techniques and materials. If we accept that premise, in 1998 over 10% of all joint replacements were performed for wear-related problems and still constitute the major clinical problem related to total hip replacement.

The 1994 NIH consensus statement on total hip replacement concluded that the major remaining issues of concern included the long-term fixation of the acetabular component, osteolysis due to wear debris, the biologic response to particles, and problems related to revision surgery. Although acetabular fixation is no longer a problem, wear and related complications continue to be the major issue affecting the longevity of total hip replacements. The bone loss associated with osteolysis can result in pelvic dissociation and instability and major segmental cortical defects in the femur. These problems require complex revision surgery, including major allografts and custom implants. Revision surgery has a less predictable outcome than primary surgery and is less likely to lead to full restoration of function.

Osteolysis, a consequence of the wear process, is a late-appearing complication. Massive lesions can develop before patients report any clinical symptoms. Continued periodic follow-up is important to ensure remedial action before catastrophic failure occurs. Young, active patients are most at risk for wear and osteolysis. Preventive and remedial efforts should focus on this patient population.

Since 1995, the scope of wear-related problems in total joint arthroplasty has expanded to include not only the local effects of debris but also systemic distribution and effects, especially in the malfunctioning implant. Although there is evidence that particulate and ionic degradation products migrate beyond the joint to remote tissues and organs, little is known about the clinical consequences. Consideration of potential systemic effects is warranted, particularly as joint replacement surgery extends to younger patient populations and prosthesis survivorship increases.

In the past 5 years, several new wear-resistant bearing surfaces for total hip replacement have been introduced either in FDA-controlled clinical studies or as implants approved for clinical use, including ceramic-ceramic, metal versus metal articulations, and elevated cross-linked ultra high molecular weight polyethylene. If long-term clinical follow-up studies prove successful, these innovations could dramatically affect the success and longevity of total hip replacement.

Relevance

The incidence of primary and total hip arthroplasty has changed little (less than 5%) since 1995. The incidence of revision surgery remains about the same, slightly greater than 10% of all hip joint arthroplasties. The basic problems related to revision surgery remain unresolved. Revision surgery is technically more difficult, and bone loss remains a challenge. Because 30% of total hip replacements are done in young and active patients, the problems associated with prosthetic failure and revision surgery are formidable. Furthermore, recent information on the distribution of wear particles and byproducts to remote organs leads to concern about potential systemic effects.

The introduction of materials with greater wear resistance could improve the longevity of total hip replacements. Considering the number of operations that have been performed in the past 20 years, however, wear-related complications such as osteolysis and periprosthetic bone resorption will remain the major issue.

Future Directions for Research

Research and development in wear-resistant materials continues to be a high priority. Clinical research designed to thoroughly evaluate the performance of new materials intended to reduce wear is essential to ascertain their efficacy and prevent the possibility of unexpected failure.

Research into the biology of the osteolytic process and potential effects of systemic distribution of wear debris is important. Long-term surveillance of relevant populations will also be valuable in understanding the nature and severity of potential systemic effects.

Given the large number of young patients with implanted devices, research efforts are needed to address wear-related complications in this patient population, including pharmacologic intervention and restoration of skeletal integrity and function. Long-term follow-up studies through regional or national registries would be of great value in ascertaining performance of implant systems.[17,18]

References

1. NIH Consensus Statement: *Total Hip Replacement.* 1994;12:1-31.
2. HCIA 1999 Inpatient View. HVI, Inc.
3. Schulte KR, Callaghan JJ, Kelly SS, Johnston RC: The outcome of Charnley total hip arthroplasty with cement after a minimum twenty year follow-up: The results of one surgeon. *J Bone Joint Surg Am* 1993;75:961-975.
4. Sullivan PM, MacKenzie JR, Callaghan JJ, Johnston R: Total hip arthroplasty with cement in patients who are less than fifty years old: A sixteen to twenty-two year follow-up study. *J Bone Joint Surg Am* 1994;76:863-869.
5. Berger RA, Kull LR, Rosenberg AG, Galante JO: Hybrid total hip arthroplasty: Seven to ten year results. *Clin Orthop* 1996;333:134-146.
6. Madey SM, Callaghan JJ, Olejniczak JP, Goetz DD, Johnston RC: Charnley total hip arthroplasty with use of improved techniques of cementing: The results after a minimum of fifteen years of follow-up. *J Bone Joint Surg Am* 1997;79:53-63.
7. Mulroy WF, Estok D, Harris WH: Total hip arthroplasty with use of so-called second-generation cementing techniques: A 15-year-average follow up study. *J Bone Joint Surg Am* 1995;77:1845-1852.
8. Sporer S, Callaghan J, Olejniczak J, Goetz D, Johnston R: The effect of surface roughness and polymethylmethacrylate precoating on the radiographic and clinical results of the Iowa hip prosthesis: A study of patients less than fifty years old. *J Bone Joint Surg Am* 1999;81:481-492.

9. Ragab AA, Kraay MJ, Goldberg VM: Clinical and radiographic outcomes of total hip arthroplasty with insertion of an anatomically designed femoral component without cement for the treatment of primary osteoarthritis: A study with a minimum of six years of follow-up. *J Bone Joint Surg Am* 1999;81:210-218.

10. Martell JM, Pierson RH, Jacobs JJ, et al: Primary total hip reconstruction with a titanium fiber-coated prosthesis inserted without cement. *J Bone Joint Surg Am* 1993;75:554-571.

11. Thanner J: The acetabular component in total hip arthroplasty: Examination of different fixation principles. *Acta Orthop Scand* 1999;286(suppl):70.

12. Tompkins GS, Jacobs JJ, Kull LR, Rosenberg AG, Galante JO: Primary total hip arthroplasty with a porous-coated acetabular component: Seven-to-ten-year results. *J Bone Joint Surg Am* 1997;79:169-176.

13. McKellop HA, Campbell P, Park SH, et al: The origin of submicron polyethylene wear debris in total hip arthroplasty. *Clin Orthop* 1995;311:3-20.

14. Urban RM, Jacobs JJ, Gilbert JL, Galante JO: Migration of corrosion products from modular hip prostheses: Particle microanalysis and histopathological findings. *J Bone Joint Surg Am* 1994;76:1345-1359.

15. Urban RM, Jacobs JJ, Gilbert JL, et al: Characterization of solid products of corrosion generated by modular-head femoral stems of different designs and materials, in Marlowe DE, Parr JE, Mayor MB (eds): *Modularity of Orthopedic Implants.* American Society for Testing Materials, 1997, pp 38-44.

16. Malchau H, Herberts P, Söderman P, Odén A: Prognosis of total hip replacement: Scientific exhibit presented at the 67th Annual Meeting, American Academy of Orthopaedic Surgeons, Orlando, FL, March 15-19, 2000.

17. Garellick G, Malchau H, Herberts P: Survival of total hip replacements: A comparison of a randomized trial and a registry. *Clin Orthop* 2000;375:157-167.

18. NIH Consensus Statement. *Total Hip Replacement* 1994;12.

What is the clinical scope of implant wear in the knee and how has it changed since 1995?

The number of primary knee arthroplasties being performed in the United States continues to increase (up 32% from 189,008 in 1993 to 249,944 in 1998).[1] Furthermore, the number of knee revisions performed in 1998 was 21,364, an increase of 5,000 over revisions in 1993. The ratio of revision knees to primary knees has remained relatively constant from 1993 to 1998.

Although the frequency of revision surgery has not changed, the failure mechanism that necessitates revision total knee arthroplasty appears to be increasingly related to polyethylene wear and osteolysis. Unfortunately, the literature contains scant information on the incidence of wear and osteolysis following total knee arthroplasty. In 1991, Rand reported that implant loosening was the major mode of failure.[2] Prior to 1993, aseptic loosening was the leading reason for revision arthroplasty at the Anderson Orthopaedic Institute (32% of 201 revision procedures), followed by osteolysis and polyethylene wear (16%). From 1993 to 1998, 15% of 214 revision procedures were performed for loosening and 41% for osteolysis and polyethylene wear.

Relevance

Little is known about the true wear rate of polyethylene with total knee arthroplasty for a number of reasons. First, no radiographic methods for measuring wear have been established. The complex geometry of knee implants makes this task far more difficult than with total hip arthroplasty. Second, activity levels are largely unreported. Tibial polyethylene wear should correlate with use, not time, yet this variable is not reported in clinical outcome studies. Finally, polyethylene wear and fatigue-related wear are adversely affected by sterilization using gamma radiation in air and oxidation on the shelf. Because almost all total knee implants inserted prior to the mid 1990s were sterilized in this fashion and their shelf lives were unknown, the impact of oxidation on wear is unclear.

Design features that can accelerate polyethylene wear have been eliminated from most current total knee implants. Thin polyethylene and high-contact stresses with flat-on-flat designs were recognized as potential accelerators of polyethylene wear with tibial components in the mid 1980s. However, these older implant designs continue to account for many cases of polyethylene wear and osteolysis.

The orthopaedic community has recognized that implant sterilization can adversely impact the wear properties of polyethyelene. Oxidative degradation occurs with prolonged shelf life of polyethylene components sterilized by gamma radiation in air. Implant manufacturers have discontinued this method of implant sterilization. In addition, many components are now packaged in oxygen-free or inert environments. However, many manufacturers continue to sell polyethylene implants gamma radiated in air, with shelf lives up to 5 years. Accordingly, shelf life is a factor when selecting a component that has been sterilized in this fashion.

Most total knee implants that have been in situ more than 5 years were sterilized with gamma radiation in air. Because information on shelf life is not readily available from some implant manufacturers, many clinical reports do not include information on either the method of sterilization or shelf life of the polyethylene. These issues become confounding variables in most clinical studies. In the one study that included pertinent shelf-life information, Bohl and associates[3] reported a significant correlation between shelf life and failure by polyethylene wear with the Synatomic (Depuy, Warsaw, IN) implant. Failure rates for wear-related phenomena were 0% with a shelf life of up to 4 years following gamma radiation in air, 11% with a shelf life of 4 to 8 years, and 21% for implants with a shelf life greater than 8 years.

In the past 2.5 years, surgeons at the Anderson Orthopaedic Clinic implanted 73 Duracon (Stryker Howmedica, Allendale, NJ) unicondylar components gamma irradiated in air with shelf lives from 4.5 to 7 years. Ten implants required revision surgery within the first 2 years for accelerated wear. Radiographic wear has been measured at almost 1 mm per year for components implanted in 1998 (shelf life over 5 years) and approaches 2 mm per year for those implanted in 1999 (shelf life over 6 years). High levels of oxidation were confirmed from measurements on the retrieved components, and the components demonstrated severe wear, embrittlement, and fracture (Fig. 1).

Modularity has been identified as a source of wear debris with total joint implants. Modular devices improve inventory availability and provide intraoperative options, but from a wear perspective modularity is a compromise. Although articular side wear has long been considered the primary source of debris, osteolysis is rarely encountered with nonmodular knee components. During the 1970s and early 1980s, polyethylene wear and osteolysis were not identified with total knee replacements that featured all polyethylene or metal-backed one-piece tibial components. In the second edition of *Surgery of the Knee*, Insall did not even index the word osteolysis.[4] Compression-molded polyethylene and sterilization by ethylene oxide in early nonmodular implants generally had excellent results without failure by wear or osteolysis.

Peters and associates[5] reported osteolysis following Synatomic total knee arthroplasty with a modular tibial component in 1992. Osteolysis subsequently was reported with a variety of modular implants.[6-8] Tibial polyethylene backside wear was first reported in a retrieval analysis study[9] and subsequently correlated with the occurrence of osteolysis.[10] These clinical find-

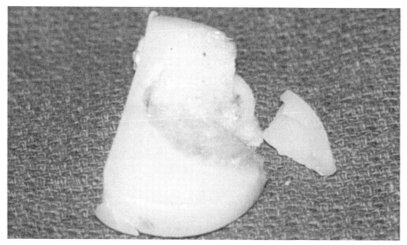

Figure 1 *Severe wear of a Duracon unicondylar component in situ 8 months. The shelf life of this component, which was gamma-irradiated in air, exceeded 5 years.*

ings led to investigations into the stability of tibial tray locking mechanisms to determine if modularity could create a secondary wear couple.[11] Fresh unimplanted modular components of 9 different designs demonstrated significant motion (> 100 μm) between the tibial baseplate and the modular polyethylene with loads of 100 N. Subsequent investigations measuring tibial tray displacement with the same implants under almost no load reported in-plane motion of 50 to 100 μm.[12]

Tibial polyethylene backside wear was reported more recently in two other studies.[13,14] A stippled pattern of polyethylene wear and scratching and burnishing of the metal baseplate have been identified (Fig. 2).[12] These investigations indicate that backside wear can produce wear debris that may be responsible for osteolysis. Furthermore, locking mechanism stability deteriorates, demonstrated by the significant displacement between the insert and baseplate in retrieved implants.

Wear with total knee arthroplasty is far more dependent on alignment and ligament balancing techniques than wear with total hip arthroplasty. Our understanding of alterations of knee kinematics with total knee arthroplasty

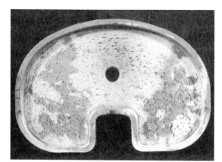

Figure 2 Left, *Stippled pattern of baseplate wear in a PFC modular component retrieved after 3.5 years in situ. Right, Underside of the matched polyethylene insert demonstrates loss of manufacturer's product identification numbers secondary to wear.*

has markedly improved,[15] but the surgical technique for total knee arthroplasty remains largely unchanged. Improvements in surgical technique with experience and teaching should reduce the frequency of component malpositioning that tends to accelerate wear.

Future Directions for Research

Clinical Research

Because the methods for sterilization of polyethylene were changed only recently, the success of newer methods in eliminating or delaying the onset of oxidative degradation is unknown. However, clinical reports of total knee arthroplasty failures should include documentation of the method of sterilization and product shelf life prior to implantation. Such information will permit us to evaluate the influence of other design and material issues that may affect polyethylene wear.

Retrieval Research

Retrieved tibial components should be examined for backside as well as articular side wear. Levels of polyethylene oxidation should be measured with retrieved polyethylene to determine if oxidation is initiated in vivo regardless of the method of sterilization.

Mechanical Testing

Knee simulator studies should examine the problem of backside wear under physiologic loading conditions, including translation, lift-off, and rotation.[16] Backside wear debris should be investigated to determine if the debris particles are morphologically different than articular side wear and to determine if significant metal debris particles are generated.

References

1. Mendenhall S: *Orthopaedic Network News.* 2000;11:1.
2. Rand JA: Revision total knee arthroplasty: Techniques and results, in Morrey BF (ed): *Joint Replacement Arthroplasty.* New York, NY, Churchill Livingstone, 1991, pp 1093-1111.
3. Bohl JR, Bohl WR, Postak PD, Greenwald AS: The effects of shelf life on clinical outcome for gamma sterilized polyethylene tibial components. *Clin Orthop* 1999;367:28-38.
4. Insall JN: *Surgery of the Knee,* ed 2. New York, NY, Churchill Livingstone, 1993.
5. Peters PC Jr, Engh GA, Dwyer KA, Vinh TN: Osteolysis after total knee arthroplasty without cement. *J Bone Joint Surg Am* 1992;74:864-876.
6. Sanchez-Sotelo J, Ordonez JM, Prats SB: Results and complications of the low contact stress knee prosthesis. *J Arthroplasty* 1999;14:815-821.
7. Robinson EJ, Mulliken BD, Bourne RB, Rorabeck CH, Alvarez C: Catastrophic osteolysis in total knee replacement: A report of 17 cases. *Clin Orthop* 1995;321:98-105.
8. Ayers DC: Polyethylene wear and osteolysis following total knee replacement. *Instr Course Lect* 1997;46:205-213.

9. Engh GA, Dwyer KA, Hanes CJ: Polyethylene wear of metal-backed tibial components in total and unicompartmental knee prosthesis. *J Bone Joint Surg Br* 1992;74:9-17.

10. Wasielewski RC, Parks N, Williams I, Surprenant H, Collier JP, Engh GA: Tibial insert undersurface as a contributing source of polyethylene wear debris. *Clin Orthop* 1997;345:53-59.

11. Parks NL, Engh GA, Topoleski LDT, Emperado J: Modular tibial insert micromotion: A concern with contemporary knee implants. *Clin Orthop* 1998;356:10-15.

12. Engh GA, Rao A, Ammeen DJ, Sychterz CJ: Abstract: Tibial baseplate wear: A major source of debris with contemporary modular knee implants. *67th Annual Meeting Proceedings*. Rosemont, IL, American Academy of Orthopaedic Surgeons, 2000, p 611.

13. Furman BD, Schmeig JJ, Bhattacharya S, Li S: Assessment of backside polyethylene wear in 3 different metal backed total knee designs. *Trans Orthop Res Soc* 1999;45:149.

14. Conditt MA, Stein J, Noble PC: Abstract: Backside wear of polyethylene modular tibial inserts. *67th Annual Meeting Proceedings*. Rosemont, IL, American Academy of Orthopaedic Surgeons, 2000, p 580.

15. Dennis DA, Komistek RD, Colwell CE Jr, et al: In vivo anteroposterior femorotibial translation of total knee arthroplasty: A multicenter analysis. *Clin Orthop* 1998;356:47-57.

16. Stiehl JB, Dennis DA, Komistek RD, Crane HS: In vivo determination of condylar lift-off and screw-home in a mobile-bearing total knee arthroplasty. *J Arthroplasty* 1999;14:293-299.

What patient-related factors contribute to implant wear?

Wear of a polyethylene bearing surface is a function of the amount and type of use and the conditions under which the bearing operates. The same principle likely applies to metal-on-metal and ceramic-on-ceramic bearings, but limited data are available.

Patients influence wear by determining the type and amount of use of the prosthetic joint. Variables such as age, gender, height, weight, body mass index, the etiology of joint arthritis, and comorbidities are easily measured and are frequently used as surrogates for the more fundamental variable, lower extremity activity.

For the majority of patients, walking is the activity that contributes most to the wear of total hip and knee prostheses.[1] An electronic digital pedometer was used to record the number of steps taken by a group of 111 nonrandomized patient volunteers with at least one total hip or total knee replacement.[2] These patients averaged 4,988 steps per day, which extrapolates to approximately 0.9 million cycles per year for a lower extremity joint. The range in activity was quite large, however, ranging from 395 to 17,718 steps per day. The most active patient walked more than 3.5 times the average. Cohorts of total joint patients from other geographic locations or cultures will likely have differences in their average activity simply due to differences in lifestyle.

Age was significantly correlated to activity ($P < 0.05$) but with a high degree of variability (standard deviation 3,040 steps per day) (Fig. 1). Patients younger than 60 years walked 30% more than those over 60. Male patients walked 28% more than female patients, and males under 60 walked 40% more than the other patients. These data indicate that individual differences in patient activity can be a substantial source of variability in polyethylene wear rates.[2]

Because some patient characteristics have shown a relationship to walking activity or wear in cohort studies, these characteristics have been used as surrogates for activity or wear. Caution should be exercised, however, when applying such trends on an individual basis. As shown in Figure 1, some patients over 70 years of age were very active, while other patients under 40 were relatively inactive. Age alone is not a salient criterion for prosthesis selection.

Not surprisingly, an inverse relationship has been demonstrated between activity and obesity. In an evaluation of 209 individuals from 22 to 82 years of age using a pedometer, body mass index was negatively associated with activity after adjusting for age, gender, and Charnley class. Patient weight

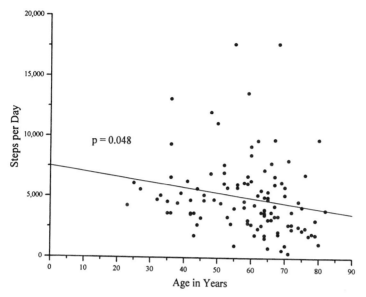

Figure 1 *Patient activity versus age. (Reproduced with permission from Schmalzried TP, Szuszczewicz ES, Northfield MR, Akizuki KH, Frankel RE, Amstutz HC: Quantitative assessment of walking activity after total hip or knee replacement. J Bone Joint Surg Am 1998;80:54-59.)*

has not shown a consistent relationship to wear. The body mass indices of the subjects with a total knee replacement were higher than those with a total hip replacement, and the activities of subjects with total knee replacements were correspondingly lower.[3]

In another study, the walking activity of 100 total joint replacement patients was measured with a pedometer and compared to the UCLA activity score and a simple visual-analog scale. Both the UCLA activity rating and the visual-analog scale rating of the investigator had a strong correlation with the pedometer data. There was, however, up to a 15-fold difference in the average steps per day for individual patients with the same UCLA score. Adjustment of the UCLA activity score for the frequency and intensity of activity, as can be done with the investigator visual-analog scale, increases the accuracy of the rating.[4]

A questionnaire was used to estimate patient walking distance in a group of 79 patients with 109 Charnley low-friction arthroplasties. Based on this assessment, an activity grade was assigned to each patient, and a significant correlation was shown between activity and acetabular component wear.[5] A strong correlation was demonstrated between another simple activity rating scale and volumetric polyethylene wear in a cohort of 80 total hip patients.[6]

A more accurate and thorough evaluation of the ambulatory activity of patients with joint replacements has been obtained using a SAM (step activity monitor), which is a microprocessor worn on the ankle that records steps in real time.[7] Comparing a pedometer to a SAM in a field trial of 33 patients revealed that the pedometer missed some lower extremity movements that could contribute to the wear of total hip and total knee prostheses. The SAM recorded 34% more cycles per day than the pedometer, but the two measures were highly correlated. The pedometer undercounts most frequently in relatively obese females.

Male gender and younger age have frequently been associated with higher polyethylene wear rates in vivo. In a radiographic study of 1,024 hips, the linear penetration of the femoral head was 37% greater in male patients compared to female patients, and 33% greater in patients under 60 years of age compared to older patients.[8] In univariate analyses, the relationships between weight, height, the etiology of arthritis, Charnley class, and polyethylene wear have been less consistently related to wear (Table 1).[5,6,9-22]

Because multiple factors affect wear, stratification of salient variables—including activity—is necessary to determine their relative effect on wear. Linear and volumetric wear in 37 hips was measured from digital images using a validated 2-dimensional, edge detection-based computer algorithm. Regression analyses were performed to evaluate the relationship between polyethylene wear and patient-related variables. Activity was measured with a SAM, and no difference was found in the number of gait cycles between the males and the females in this cohort. Univariate regression analysis indicated that the strongest factor influencing wear was male gender, followed by height and weight (both of which were highly correlated to male gender). Higher walking speed (steps per minute) was also positively correlated to wear. Multivariate regression analysis revealed that male gender had a strong effect on wear, independent of other variables such as height, weight, and activity. Males had a higher average walking speed and spent more time walking fast than did females. The effect that male gender had on wear may be attributable to differences in behavior (including the types or patterns of activity), anatomy (as it affects the stresses on the joint), physiology (such as the lubricant properties of synovial fluid), or a combination of these type factors. Wear was highly correlated to joint use. A visual-analog scale activity rating of the patient by the investigator was significantly related to wear.[4] Wear is a function of use, not time.[22]

The assessment of polyethylene wear in total knee replacement is substantially more difficult than for total hip replacement because the geometries of the bearing surfaces of total knees are more complex and the radiographic evaluations less precise. For these reasons, there is a paucity of information on wear in total knees. Surrogate measures, such as surface damage and reoperation rates, have been more commonly reported.

Surgical technique, implant design, and material limitations further complicate the assessment of polyethylene wear in knee arthroplasty. Examples of confounding variables include malalignment and instability, low conformity, thin polyethylene, oxidation, and insecure locking mechanisms for modular components. These factors can overshadow the contribution of patient-related factors to wear and osteolysis and make comparisons within and between studies very difficult.

As summarized in Table 2,[23-31] none of the studies on knee replacements evaluated patient activity as a factor in polyethylene wear or osteolysis in total knee replacements. Similar to total hip replacement, young age and

Table 1 Patient-related factors and polyethylene wear in total hip replacement

Study	No. Hips	No. Designs	Age	Male Gender	Weight	Height	Diagnosis	Activity
Charnley et al (1973)	106	1	-	-	NO	-	-	NO
Charnley et al (1975)	135	1	-	-	NO	-	-	NO
Griffith et al (1978)	491	1	YES	YES	NO	-	-	NO
Rimnac et al (1988)	10	1	YES	-	NO	-	-	-
Livermore et al (1990)	385	3	-	-	YES	-	-	-
Kim et al (1993)	116	1	YES	NO	YES	-	NO	-
Feller et al (1994)	109	1	NO	NO	NO	NO	NO	YES
Callaghan et al (1995)	210	5	YES	YES	-	-	-	-
Nashed et al (1995)	175	4	YES	YES	NO	-	-	-
Xenos et al (1995)	100	1	YES	NO	NO	NO	-	-
Zicat et al (1995)	137	2	YES	-	-	-	NO	NO
Devane et al (1997)	80	1	YES	NO	NO	-	-	YES
Jasty et al (1997)	128	9	NO	NO	-	-	-	-
Schmalzried et al (1997)	1,024		YES	YES	NO	NO	NO	NO
Shih et al (1997)	240		YES	NO	NO	-	NO	-
Perez et al (1998)	27	2	YES	NO	Negative correlation	-	-	-
Schmalzried et al (2000)	37		YES	YES	YES	YES	NO	YES

male gender are associated with higher polyethylene damage scores, component failure, and osteolysis. In total knee replacement, patient weight shows a frequent association with polyethylene damage, component failure, and osteolysis. Activity assessments of patients with total knee replacements have demonstrated high functional capabilities in a cohort of relatively young patients with a mean age of 51 years (range, 22 to 55). Unfortunately,

Table 2 Patient-related factors and polyethylene wear in total knee replacement

Study	No. Knees	No. Designs	Age	Gender	Weight	Height	Diagnosis	Wear Parameter
Hood et al (1983)	48	1	-	-	YES	-	-	Damage score
Kilgus et al (1991)	176	1	NO	YES	YES	-	-	Failed components
Engh et al (1992)	86	17	-	NO	NO	NO	NO	Damage score
Cadambi et al (1993)	271	2	YES	YES	YES	-	YES	Osteolysis
Tsao et al (1993)	487	1	YES	NO	YES	NO	NO	Damage score
Kim et al (1995)	60	1	NO	NO	NO	-	-	Osteolysis
Tanner et al (1995)	29	4	NO	NO	NO	-	-	Damage score
Engh et al (2000)	48	5	YES	YES	-	YES	-	Damage score

wear of these polyethylene components could not be accurately measured on conventional radiographs.[32]

Relevance

Radiographic and retrieval analyses of polyethylene generally report wear as a function of time. The limitations of this methodology need to be appreciated because wear is a function of use. In cohort analyses, the results should be stratified according to recognized covariables that include patient gender, age, height, weight, and body mass index. Gender and age may be practical surrogates for activity in cohort studies, but have little value in individual cases. Including any measure of patient activity, such as simply ranking activity from grade 1 (inactive) to grade 5 (very active), would enhance studies of patient-related factors and wear in total joint prostheses. In randomized, prospective studies, a measure of joint use is needed to assess the distribution of this critical variable among the study groups.

Future Directions for Research

Some differences in patient activity may exist as a function of geographic location and lifestyle. To understand better the range of demands on total joint prostheses, quantitative activity data are needed on patients from a variety of regions, races, and cultures. These data can assist in the development, validation, and implementation of simple activity rating scales for use in routine clinical evaluations. Further investigations into possible gender-related differences in polyethylene wear are also needed. It is also likely that the type and intensity of activity affect wear, and more detailed investigations of patient activity are needed. Activity data from such clinical studies can be used to direct and improve wear simulations. Finally, techniques must be developed to improve quantitative assessment of polyethylene wear in total knee replacements, in both radiographic and retrieval analyses.

References

1. Seedhom BB, Wallbridge NC: Walking activities and wear of prostheses. *Ann Rheum Dis* 1985;44:838-843.
2. Schmalzried TP, Szuszczewicz ES, Northfield MR, Akizuki KH, Frankel RE, Amstutz HC: Quantitative assessment of walking activity after total hip or knee replacement. *J Bone Joint Surg Am* 1998;80:54-59.
3. McClung CD, Zahiri CA, Higa JK, Amstutz HC, Schmalzried TP: Relationship between body mass index and activity in hip or knee arthroplasty patients. *J Orthop Res 2000*;18:35-39.
4. Zahiri CA, Schmalzried TP, Amstutz HC: Assessing activity in joint replacement patients. *J Arthroplasty* 1998;13:890-895.
5. Feller JA, Kay PR, Hodgkinson, Wroblewski BM: Activity and socket wear in the Charnley low-friction arthroplasty. *J Arthroplasty* 1994;9:341-345.
6. Devane PA, Horne JG, Martin K, Coldham G, Krause B: Three-dimensional polyethylene wear of a press-fit titanium prosthesis: Factors influencing generation of polyethylene debris. *J Arthroplasty* 1997;12:256-266.
7. Shepherd EF, Toloza E, McClung CD, and Schmalzried T: Step activity monitor (SAM): Increased accuracy in quantifying ambulatory activity. *J Orthop Res 1999*;17:703-708.
8. Schmalzried TP, Dorey FJ, McClung CD, et al: Factors contributing to the variability of short-term linear penetration rates in total hip replacement. *Orthop Trans 1997*;21:228-229.
9. Charnley J, Cupic Z: The nine and ten year results of the low-friction arthroplasty of the hip. *Clin Orthop* 1973;95:9-25.
10. Charnley J, Halley DK: Rate of wear in total hip replacement. *Clin Orthop* 1975;112:170-179.
11. Griffith MJ, Seidenstein MK, Williams D, Charnley J: Socket wear in Charnley low friction arthroplasty of the hip. *Clin Orthop* 1978;137:37-47.
12. Rimnac CM, Wilson PD, Fuchs MD, Wright TM: Acetabular cup wear in total hip arthroplasty. *Orthop Clin North Am* 1988;19:631-636.
13. Livermore J, Ilstrup D, Morrey B: Effect of femoral head size on wear of the polyethylene acetabular component. *J Bone Joint Surg Am* 1990;72:518-528.
14. Kim Y-H, Kim VEM: Uncemented porous-coated anatomic total hip replacement. Results at six years in a consecutive series. *J Bone Joint Surg Br* 1993;75:6-13.
15. Callaghan JJ, Pedersen DR, Olejniczak JP, Goetz DD, Johnston RC: Radiographic measurement of wear in 5 cohorts of patients observed for 5 to 22 years. *Clin Orthop* 1995;317:14-18.
16. Nashed RS, Becker DA, Gustilo RB: Are cementless acetabular components the cause of excess wear and osteolysis in the total hip arthroplasty? *Clin Orthop* 1995;317:19-28.
17. Xenos JS, Hopkinson WJ, Callaghan JJ, Heekin RD, Savory CG: Osteolysis around an uncemented cobalt chrome total hip arthroplasty. *Clin Orthop* 1995;317:29-36.
18. Zicat B, Engh CA, Gokcen E: Patterns of osteolysis around total hip components inserted with and without cement. *J Bone Joint Surg Am* 1995;77:432-439.

19. Jasty M, Goetz DD, Bragdon CR, et al: Wear of polyethylene acetabular components in total hip arthroplasty: An analysis of 128 components retrieved at autopsy or revision operations. *J Bone Joint Surg Am* 1997;79:349-358.

20. Shih C-H, Lee P-C, Chen J-H, et al: Measurement of polyethylene wear in cementless total hip arthroplasty. *J Bone Joint Surg Br* 1997;79:361-365.

21. Perez RE, Rodriguez JA, Deshmukh RG, Ranawat CS: Polyethylene wear and periprosthetic osteolysis in metal-backed acetabular components with cylindrical liners. *J Arthroplasty* 1998;13:1-7.

22. Schmalzried TP, Shepherd EF, Dorey FJ: The 2000 John Charnley Award: Wear is a function of use, not time. *Clin Orthop* 2000;381:36-46.

23. Hood RW, Wright TM, Burstein AH: Retrieval analysis of total knee prostheses: a method and its application to 48 total condylar prostheses. *J Biomed Mater Res* 1983;17:829-842.

24. Wright TM, Bartel DL: The problem of surface damage in polyethylene total knee components. *Clin Orthop* 1986;205:67-74.

25. Kilgus DJ, Moreland JR, Finerman GAM, Funahashi TT, Tipton JS: Catastrophic wear of tibial polyethylene inserts. *Clin Orthop* 1991;273:223-231.

26. Engh GA, Dwyer KA, Hanes CK: Polyethylene wear of metal-backed tibial components in total and unicompartimental knee prostheses. *J Bone Joint Surg Br* 1992;74:9-17.

27. Cadambi A, Engh GA, Dwyer KA, Vinh TN: Osteolysis of the distal femur after total knee arthroplasty. *J Arthroplasty* 1994;9:579-594.

28. Tsao A, Mintz L, McRay CR, Stulberg D, Wright T: Failure of the porous-coated anatomic prosthesis in total knee arthroplasty due to severe polyethylene wear. *J Bone Joint Surg Am* 1993;75:19-26.

29. Kim Y-H, Oh J-G, Oh S-H: Osteolysis around cementless porous-coated anatomic knee prosthesis. *J Bone Joint Surg Br* 1995;77:236-241.

30. Tanner MG, Whiteside LA, White SE: Effect of polyethylene quality on wear in total knee arthroplasty. *Clin Orthop* 1995;317:83-88.

31. Engh GA, Koralewicz LM, Pereles TR: Clinical results of modular polyethylene insert exchange with retention of total knee arthroplasty components. *J Bone Joint Surg Am* 2000;82:516-523.

32. Diduch DR, Insall JN, Scott WN, Scuderi GR, Font-Rodriguez D: Total knee replacement in young, active patients: Long-term follow-up and functional outcome. *J Bone Joint Surg Am* 1997;79:575-581.

What surgical-related factors contribute to implant wear?

Wear is affected by many variables, including design and material properties of the component, the response of the host to particles, and the mechanical environment of the joint. A number of studies have addressed the clinical issues of patient selectivity and activity as an important underlying cause of enhanced wear.[1] Surgical technique is equally important, but has not been given the same emphasis in the literature as other issues that affect the incidence and progression of wear in total hip and knee replacement components.

Several studies have addressed the relationship of alignment of the acetabular component to polyethylene wear. An angle of inclination of greater than 50° appears to predispose a component to increased polyethylene wear and subsequent pelvic osteolysis.[2-9] The vertical position of the cup subjects the component to increased peripheral wear of the liner, a region with reduced polyethylene thickness in some designs. When the acetabular component was placed in a more anatomically horizontal position of inclination, wear and subsequent pelvic osteolysis was significantly reduced and excellent rim fixation was maintained. Further anatomic restoration of the hip center of rotation and offset has been reported to decrease the progression of wear. Impingement between acetabular and femoral components produces wear debris and also results in abnormal loading on the edge of the implant.[3] One study associated increased impingement with excessive cup anteversion combined with posterior positioning of an extended liner rim and use of a femoral component with a relatively small head-to-neck diameter ratio.[3] The linear wear rate of components without impingement was significantly lower than those cups that demonstrated impingement.

There appears to be an increase in polyethylene wear when a cup is cemented with an incomplete or asymmetrical cement mantle. An unsatisfactory cement technique used for the femoral component also can result in increased wear and resulting osteolysis. When a cementless socket is implanted with inadequate bony support or incomplete reaming of the acetabulum, creating gaps between the component and the underlying bone, excessive wear and osteolysis may result. Because third-body debris can accelerate wear in hip and knee replacements by abrasive mechanisms, removal of loose cement and bone particles during surgery is important.[10-12] Studies suggest that debris generated in vivo from loose acetabular or femoral components can migrate into the space between the femoral head and acetabular polyethylene, thus significantly enhancing wear.[1] Accelerated polyethylene wear due to third-body debris can be caused by the use of braided cables. One study demonstrated significantly more acetabular wear in hips repaired with braided cables than in hips repaired with monofilament wire.[1] These findings concerning third-body particles underscore the importance of removing as much surgically induced debris as possible. Although improved spatial orientation of the components of total hip replacements might decrease force transmission across the joint and subsequent polyethylene wear, few studies have defined this relationship.

The wear of polyethylene tibial and patellar components has been closely correlated with surgical technique in total knee replacement. Wasielewski and associates correlated increased polyethylene wear with inadequate medial release, excessive posterior slope of the tibial component, elevation of the jointline, and patellar subluxation.[13] Malrotation in total knee arthroplasty has also been implicated in increased posteromedial polyethylene wear.[14] Uncorrected knee alignment and failure to balance the soft tissues will subject the total knee replacement to accelerated wear. If the posterior cruciate ligament is retained, it must be balanced; if not, increased posterior tibial polyethylene wear may occur, resulting in arthroplasty failure.[15,16] Significant polyethylene wear is also associated with progressive femoral-tibial subluxation as a result of postoperative limb malalignment or excessive varus positioning of the tibial component.[17,18] Malalignment of the patellar component and elevation of the jointline more than 4 mm in a posterior cruciate-retaining or 8 mm in a posterior cruciate-substituting knee replacement subject the patellar component to excessive wear and failure.[19] Surgical technique is critical in determining the extent of wear in total knee replacement; no prosthetic design can substitute for excellent surgical technique and proper component alignment. Although surgical technique is an important determinant of component wear, patient selection may be equally important. A significant number of studies suggest that younger, heavier patients are predisposed to polyethylene wear.[1,8,20]

Relevance

Surgical technique can have a critical direct impact on the prevention and reduction of wear of joint replacement components. Meticulous preoperative planning and surgical technique, together with improved instrumentation and implant components, will reduce wear and enhance the longevity of a replacement.

Future Directions for Research

Additional prospective studies of the relationship between component alignment and wear in both total hip and total knee replacements should be initiated. Wear patterns observed in retrieved implants should be directly correlated to surgical technique. Implant retrieval registries must have available complete clinical data to develop meaningful correlations of wear to surgical variables. Sophisticated models reconstructing the spatial positioning of components should be directly correlated to wear patterns seen in autopsy retrievals. Radiostereometric analysis measurements should be used to define implant positioning and correlate this to specific linear and volumetric wear. Studies should be implemented to determine whether the development of more robust designs for total joint replacements will help reduce wear associated with less than ideal surgical technique.

References

1. Schmalzried TP, Callaghan JJ: Current concepts review: Wear in total hip and knee replacements. *J Bone Joint Surg Am*; 1999;81:115-136.
2. Kennedy JG, Rogers WB, Soffee KE, et al: Effect of acetabular component orientation on recurrent dislocation, pelvic osteolysis, polyethylene wear, and component migration. *J Arthroplasty* 1998;13:530-534.
3. Yamaguchi M, Akisue T, Bauer TW, et al: The spatial location of impingement in total hip arthroplasty. *J Arthroplasty* 2000;15:305-313.
4. Schmalzried TP, Guttermann D, Grecula M, et al: The relationship between the design, position, and articular wear of acetabular components inserted without cement and the development of pelvic osteolysis. *J Bone Joint Surg Am* 1994;76:677-688.
5. Bono JV, Sanford L, Toussaint JT: Severe polyethylene wear in total hip arthroplasty: Observations from retrieved AML PLUS hip implants with an ACS polyethylene liner. *J Arthroplasty* 1994;9:119-125.
6. Kobayashi S, Eftekhar NS, Terayama K, et al: Risk factors affecting radiological failure of the socket in primary Charnley low friction arthroplasty: A 10- to 20-year followup study. *Clin Orthop* 1994;306:84-96.
7. McCoy TH, Salvati EA, Ranawat CS, et al: A fifteen-year follow-up study of one hundred Charnley low-friction arthroplasties. *Orthop Clin North Am* 1988;19:467-476.
8. Griffith MJ, Seidenstein MK, Williams D, et al: Socket wear in Charnley low friction arthroplasty of the hip. *Clin Orthop* 1978;137:37-47.
9. Rimnac CM, Wilson PD Jr, Fuchs MD, et al: Acetabular cup wear in total hip arthroplasty. *Orthop Clin North Am* 1988;19:631-636.
10. Helmers S, Sharkey PF, McGuigan FX: Efficacy of irrigation for removal of particulate debris after cemented total knee arthroplasty. *J Arthroplasty* 1999;14:549-552.
11. Bobyn JD, Tanzer M, Krygier JJ, et al: Concerns with modularity in total hip arthroplasty. *Clin Orthop* 1994;298:27-36.
12. Davidson JA, Poggie RA, Mishra AK: Abrasive wear of ceramic, metal, and UHMW-PE bearing surfaces from third-body bone, PMMA bone cement, and titanium debris. *Biomed Mater Eng* 1994;4:213-219.
13 .Wasielewski RC, Galante JO, Leighty RM, et al: Wear patterns on retrieved polyethylene tibial inserts and their relationship to technical considerations during total knee arthroplasty. *Clin Orthop* 1994;299:31-43.
14. Eckoff DG, Metzger RD, Vandewalle MV: Malrotation associated with implant alignment technique in total knee arthroplasty. *Clin Orthop* 1995;321:28-31.
15. Swany MR, Scott RD: Posterior polyethylene wear in posterior cruciate ligament-retaining total knee arthroplasty: A case study. *J Arthroplasty* 1993;8:439-466.
16. Benjamin JB, Szivek J, Dersam G, et al: Linear and volumetric wear of tibial polyethylene retrievals from PCL retaining knee arthroplasties, in Proceedings of the 32nd Annual Meeting of the Hip Society. Belmont, CA, *Hip Society* 2000, p 90.

17. Feng EL, Stulberg SD, Wixson RL: Progressive subluxation and polyethylene wear in total knee replacements with flat articular surfaces. *Clin Orthop* 1994;299:60-71.
18. D'Lima D, Hermida JC, Chen PC, et al: Polyethylene contact stresses and knee wear: A finite element model and in vitro wear simulation, in *Proceedings of the 32nd Annual Meeting of the Hip Society*. Belmont, CA, Hip Society, 2000, p 86.
19. Goldberg VM, Kraay MJ, Buly RI, et al: Cementless total knee arthroplasty for osteoarthritis. A 5 to 9 year functional and radiological outcome study. *Orthop Trans* 1994;18:1003-1004.
20. Tsao A, Mintz L, McRae CR, et al: Failure of the porous-coated anatomic prosthesis in total knee arthroplasty due to severe polyethylene wear. *J Bone Joint Surg Am* 1993;75:19-26.

How should wear-related implant surveillance be carried out and what methods are indicated to diagnose wear-related problems?

Attempts to minimize bearing surface wear of total joint arthroplasty by using carbon fiber-reinforced polyethylene and Hylamer have failed to demonstrate a reduction in wear.[1] To avoid the mistakes of the past, wear-related surveillance must be implemented as new forms of polyethylene and hard bearing surfaces are introduced for total joint arthroplasty.

The tools available to an orthopaedic surgeon for the diagnosis of wear include patient history, physical examination, and various diagnostic tests. A significant problem arises, however, when applying these general medical diagnostic principles to wear of joint replacements. Except in catastrophic cases, implant wear may not manifest clinical symptoms for many years. Although wear occurs in all orthopaedic implants, it is well tolerated by most patients, particularly the elderly and less active. The challenge is to determine whether the material functions in vivo as well as it does in pre-clinical in vitro tests.

Implant failure resulting from catastrophic wear is easily diagnosed. In cases of implant breakage or loosening associated with the loss of structural bone support from massive osteolysis, the patient is in pain and cannot bear weight. More subtle findings related to wear include unexplained effusions. Occasionally the patient may have a mass in the region of the implant that usually can be delineated by radiographic techniques. Vague abdominal and groin complaints can also be the first signs of a nonpalpable large intrapelvic polyethylene-induced granuloma that can be identified by computed tomography. Gross instability from catastrophic wear can be diagnosed by clinical examination and routine radiographs. Complete wear of polyethylene inserts of metal-backed components can be diagnosed audibly by a screeching sound from the implant site that is associated with metal-on-metal articulation. The radiographs in these cases may show a metal arthrogram outlining the joint capsule or effective joint space.

The most widely used diagnostic tool for the surveillance of wear and osteolysis is serial radiographic evaluation.[2-7] Although gross wear of components (eg, 2 to 3 mm of linear penetration) and large areas of osteolysis can be detected radiographically, radiographs generally underestimate the

24

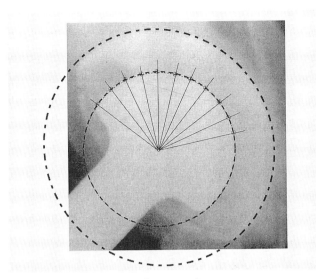

Figure 1 *Edge-detection technique for measuring wear.*

amount of osteolysis. Multiple radiographic views, including obturator and iliac oblique views in the hip and fluoroscopic spot films in the knee, can be helpful in identifying osteolysis and other interface changes.

Small amounts of femoral head penetration into the acetabular component are difficult to measure. Further complicating wear measurements is

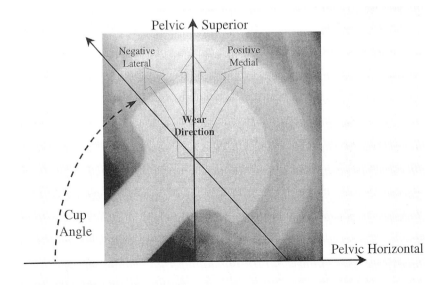

Figure 2 *Linear head penetration as well as direction of wear can be determined with the edge-detection technique.*

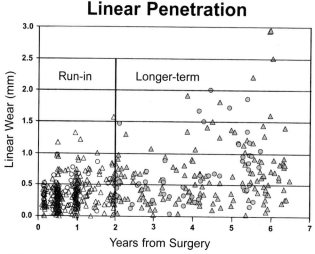

Figure 3 *Linear penetration is greater in the first 1 to 2 years.*

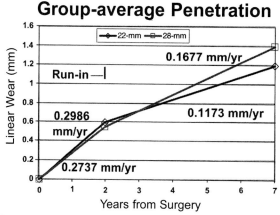

Figure 4 *Wear rates are greater in the first 2 years after surgery. The 28-mm head shows less wear than the 22-mm head in the first 2 years, but more wear after 3 years.*

the inability to standardize radiographs.[8-11] Within the realm of each clinician's work environment, attempts to standardize radiographic examination should be encouraged (eg, the position of the limbs and the tube-to-plate distance).

Manual methods of measuring head penetration, including the concentric templating technique of Livermore and associates,[3] have provided useful information regarding wear. However, the inter- and intraobserver variability is still a potential source of error.[2] Newer two- and three-dimensional computational techniques have been developed to determine the three-dimensional pattern and the volumetric amount of wear.[12-22] To measure head penetration more accurately and precisely and to avoid the problems with

Frequency Distributions

22-mm wear rates

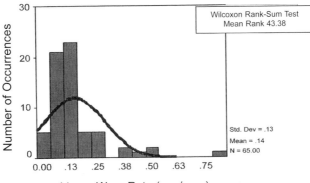

Figure 5 *A Gaussian distribution curve is superimposed on the distribution of wear rates in cohorts of patients with 22-mm heads. Note that more patients have small wear rates, with only a few high-wear outliers; the rates do not follow a Gaussian distribution.*

Frequency Distributions

28-mm wear rates

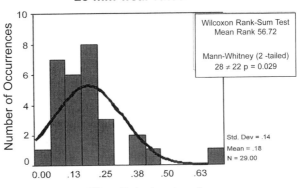

Figure 6 *A Gaussian distribution curve is superimposed on the distribution of wear rates in cohorts of patients with 28-mm heads. Like the patients with 22-mm heads, the wear rates do not follow a Gaussian distribution.*

intra- and interobserver variability, edge-detection techniques have been applied to digitally stored radiographs (Figs. 1 and 2). Radiostereometric analysis also has been used to evaluate wear in small series of patients.[23]

In addition to providing an accurate measurement of wear, it is also important to document wear over time. Three factors are important in this regard. Because more femoral head penetration is noted in the first 1 to 2 years following arthroplasty (the bedding-in period) than in subsequent years (Fig. 3), a ramped pattern (Fig. 4) more accurately depicts the wear rates demonstrated in studies of serial radiographs.[15,17] The second factor is the nonGaussian distribution of wear in large series of patients;[15] generally a large percentage of cases have minimal wear, and a small percentage have more extensive wear (Figs. 5 and 6). Finally, wear is a factor of prosthetic use rather than time.[25,26] Patient activity is a major factor in differentiating wear rates. All of these factors are critical when surveillance is employed to study and diagnose wear-related problems following total joint arthroplasty.

Most of the investigational work on evaluation of in vivo wear has been performed on total hip arthroplasty patients. Evaluation of in vivo wear following total knee replacement has been limited to implant retrieval. Computational techniques to evaluate wear on plain film radiographs are being developed; their accuracy and precision will need to be determined. In addition, ultrasound technology is being investigated as a tool for measuring knee replacement wear in vivo.[27] The geometry of joint implant surfaces, other than the hip, will continue to complicate the measurement of in vivo wear.

Relevance

The prompt and accurate detection of accelerated wear is critical when evaluating patients with a total joint arthroplasty and when evaluating a new bearing surface. Accurate, early detection of wear can enable the surgeon and implant engineer to evaluate the clinical performance of new design concepts and materials. Limiting clinical trials of new components and emphasizing the importance of ongoing wear studies may prevent the widespread use of suboptimal designs and materials. Postmarketing surveillance of wear in a new design should combine limited radiographic studies of two- and three-dimensional wear and radiostereometric analysis, large cohorts of patients from multiple surgeons, more limited studies of randomized patients and bilateral cases, and reports of all early and late failures with retrieval analysis whenever possible. All patients should be followed with serial radiographs throughout their lifetime to ensure that wear problems are diagnosed in a timely manner. Shorter follow-up intervals should be used for the higher risk (younger and more active) patients.

Future Directions for Research

Research should be directed toward the continued development and validation of diagnostic techniques for early detection and quantification of wear (eg, edge-detection techniques, three-dimensional computation, radiostereometric analysis, and newer computed tomographic imaging techniques) and its sequela (ie, osteolysis). Simple and inexpensive automated techniques that can be easily implemented in clinical practice should be developed and used in large series of patients in whom new materials and prosthetic designs have been introduced. Wear rate thresholds that might indicate osteolysis risk should be investigated, as well as potential biochemical markers that differentiate patients prone to develop osteolysis. Mechanisms and design characteristics responsible for accelerated wear should be identified and eliminated to minimize wear. Methods to evaluate wear in knee replacement and other joints should be explored. Standardized statistically valid methods, to present longitudinal wear data of individuals and cohort groups, are necessary to compare data from various investigational groups with different implant designs.

References

1. Livingston BJ, Chmell MJ, Spector M, Poss R: Complications of total hip arthroplasty associated with the use of an acetabular component with a Hylamer liner. *J Bone Joint Surg Am* 1997;79:1529-1538.
2. Callaghan JJ, Pedersen DR, Olejniczak JP, et al: Radiographic measurement of wear in 5 cohorts of patients observed for 5 to 22 years. *Clin Orthop* 1995;317:14-18.
3. Livermore J, Ilstrup D, Morrey B: Effect of femoral head size on wear of the polyethylene acetabular component. *J Bone Joint Surg Am* 1990;72:518-528.
4. Dowd JE, Sychterz CJ, Young AM, Engh CA: Characterization of long-term femoral-head-penetration rates: Association with and prediction of osteolysis. *J Bone Joint Surg Am* 2000;82:1102-1107.
5. Ilchmann T, Markovic L, Joshi A, Hardinge K, Murphy J, Wingstrand H: Migration and wear of long-term successful Charnley total hip replacements. *J Bone Joint Surg Br* 1998;80:377-381.
6. Devane PA, Horne JG: Assessment of polyethylene wear in total hip replacement. *Clin Orthop* 1999;369:59-72.
7. Hardinge K, Porter ML, Jones PR, et al: Measurement of hip prostheses using image analysis: The maxima hip technique. *J Bone Joint Surg Br* 1991;73:724-728.
8. Clarke JC, Black K, Rennie C, et al: Can wear in total hip arthroplasties be assessed from radiographs? *Clin Orthop* 1976;121:126-142.
9. Martell JM, Leopold SS, Liu X: The effect of joint loading on acetabular wear measurement in total hip arthroplasty. *J Arthroplasty* 2000;15:512-518.
10. Moore KD, Barrack RL, Sychterz CJ, Sawhney J, Yang AM, Engh CA: The effect of weight-bearing on the radiographic measurement of the position of the femoral head after total hip arthroplasty. *J Bone Joint Surg Am* 2000;82:62-69.
11. Smith PN, Ling RS, Taylor R: The influence of weight-bearing on the measurement of polyethylene wear in THA. *J Bone Joint Surg Br* 1999;81:259-265.
12. Devane PA, Bourne RB, Rorabeck CH, et al: Measurement of polyethylene wear in metal-backed acetabular cups: I. Three dimensional technique. *Clin Orthop* 1995;319:303-316.
13. Devane PA, Horne JG, Martin K, Coldham G, Krause B: Three-dimensional polyethylene wear of a press-fit titanium prosthesis: Factors influencing generation of polyethylene debris. *J Arthroplasty* 1997;12:256-266.
14. Martell JM, Berdia S: Determination of polyethylene wear in total hip replacements with use of digital radiographs. *J Bone Joint Surg Am* 1997;79:1635-1641.
15. Pedersen DR, Brown TD, Hillis SL, Callaghan JJ: Prediction of long-term polyethylene wear in total hip arthroplasty, based on early wear measurements made using digital image analysis. *J Orthop Res* 1998;16:557-563.
16. Sychterz CJ, Engh CA Jr, Shah N, Engh CA Sr: Radiographic evaluation of penetration by the femoral head into the polyethylene liner over time. *J Bone Joint Surg Am* 1997;79:1040-1046.

17. Sychterz CJ, Engh CA Jr., Yang AM, Engh CA: Analysis of temporal wear patterns of porous-coated acetabular components: distinguishing between true wear and so-called bedding-in. *J Bone Joint Surg Am* 1999;81:821-830.
18. Sychterz CJ, Yang AM, McAuley JP, Engh CA: Two-dimensional versus three-dimensional radiographic measurements of polyethylene wear. *Clin Orthop* 1999;365:117-123.
19. Urquhart AG, D'Lima DD, Venn-Watson E, Colwell CW Jr, Walker RH: Polyethylene wear after total hip arthroplasty: The effect of a modular femoral head with an extended flange-reinforced neck. *J Bone Joint Surg Am* 1998;80:1641-1647.
20. Sochart DH: Relationship of acetabular wear to osteolysis and loosening in total hip arthroplasty. *Clin Orthop* 1999;363:135-150.
21. Wroblewski BM: Direction and rate of socket wear in Charnley low-friction arthroplasty. *J Bone Joint Surg Br* 1985;67:757-761.
22. Yamaguchi M, Bauer TW, Hasimoto Y: Three-dimensional analysis of multiple wear vectors in retrieved acetabular cups. *Joint Bone Joint Surg Am* 1997;79:1539-1544.
23. Ohlin A, Selvik G: Socket wear assessment: A comparison of three different radiographic methods. *J Arthroplasty* 1993;8:427-431.
24. Franzen H, Mjobert B: Wear and loosening of the hip prosthesis: A roentgen stereophotogrammetric 3 year study of 14 cases. *Acta Orthop Scand* 1990;61:499-501.
25. Schmalzried TP, Szuszczewicz ES, Northfield MR, et al: Quantitative assessment of walking activity after total hip or knee replacement. *J Bone Joint Surg Am* 1998;80:54-59.
26. Zahiri CA, Schmalzried TP, Szuszczewicz ES, Amstutz HC: Assessing activity in joint replacement patients. *J Arthroplasty* 1998;13:890-895.
27. Yashar AA, Adler RS, Grady-Benson JC, Matthew LS, Freiberg AA: An ultrasound method to evaluate polyethylene component wear in total knee replacement arthroplasty. *Am J Orthop* 1996;25:702-704.

What are the systemic consequences of wear debris clinically?

In the vast majority of patients, permanent orthopaedic implants are well tolerated, ie they are biocompatible. In the long term, however, permanent orthopaedic implants may be associated with adverse local and remote tissue responses, mediated by the degradation products of the implant materials. These products may be present as particulate wear and corrosion debris; colloidal organometallic complexes, specifically or nonspecifically bound; free metallic ions; and inorganic metal salts/oxides; or may be sequestered in an organic storage form such as hemosiderin. Much of the focus on the long-term biocompatibility of implant materials has centered on metallic components because of their tendency to undergo electrochemical corrosion and form (at least transiently) chemically active degradation products.

Concern over the release and distribution of metallic degradation products is justified by the known potential toxicities of some of the elements used in modern orthopaedic implant alloys—titanium, aluminum, vanadium, cobalt, chromium, and nickel. Toxicity may be caused by metabolic alterations; alterations in host/parasite interactions; immunologic interactions of metal moieties due to their ability to act as haptens (specific immunologic activation), antichemotactic agents (nonspecific immunologic suppression), or lymphocyte toxins; or chemical carcinogenesis. At this time, the association of metal release from orthopaedic implants with any metabolic, bacteriologic, immunologic, or carcinogenic toxicity remains conjectural because cause and effect have not been established in humans. This may be attributed to the difficulty of observation; most symptoms of systemic and remote toxicity can be expected to occur in any patient population, not just those with implants.

Metal Ion Release

Both implants and the wear debris they generate may release chemically active metal ions. While these ions may remain in local tissues, metal ions may also bind to protein moieties that are then transported through the bloodstream and lymphatics to remote organs. Broad reviews of the toxicology of the elements used in orthopaedic metal alloys (Ti, Al, V, Co, Cr, and Ni) are available in the literature.[1-6] However, toxicities generally apply to soluble forms of these elements, and may not apply to the chemical species that result from the degradation of prosthetic implants.

Several studies have demonstrated chronic elevations in serum and urine cobalt and chromium following total joint replacement.[7-9] Transient elevations of urine and serum nickel also have been noted immediately following surgery.[7] This hypernickelemia/hypernickeluria may be unrelated to the implant itself because there is a very small percentage of nickel within these implant alloys. Rather, the elevations may be related to the use of stainless steel surgical instruments, which contain a high percentage of nickel in the alloy, or to metabolic changes associated with the surgery itself. Chronic elevations in serum titanium concentrations also have been reported in subjects with total joint replacements with titanium-containing components (Fig. 1).[9,10] Although serum and urine vanadium concentrations were not elevated in patients with total joint replacements, this is partially due to the technical difficulty associated with measuring the minute concentrations present in serum.[10]

Few studies have been conducted examining metal levels in remote anatomic sites. Metal content analyses of homogenates of remote organs and tissue obtained postmortem from subjects with cobalt-base alloy total joint components indicated that significant increases in cobalt and chromium concentrations occurred in the heart, liver, kidney, spleen, and lymphatic tissue.[11] Similarly, patients with titanium-base alloy implants have demonstrated elevated titanium levels in the spleen.[12] Spleen aluminum and liver titanium

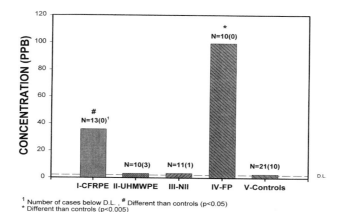

Figure 1 Bar graph of the mean concentrations of titanium in the serum, in nanograms per milliliter (parts per billion) in 4 subject groups with titanium-containing total knee replacement and a group of controls without implants. Note dramatic elevations in serum titanium in patients with failed patellar components (FP). PPB = parts per billion; DL= detection limit; CFRPE = carbon fiber reinforced polyethylene; NII = nitrogen ion-implanted. (Reproduced with permission from Jacobs JJ, Goodman SB, Sumner DR, Hallab NJ: Biologic response to orthopaedic implants, in Buckwalter JA, Einhorn TA, Simon SR (eds): Orthopaedic Basic Science: Biology and Biomechanics of the Musculoskeletal System, ed 2. Rosemont, IL, American Academy of Orthopaedic Surgeons, 2000, pp 401-426.)

levels also can be markedly elevated in some patients with failed titanium-alloy implants.[12] Spectroscopic analysis of homogenated tissue digests can be a very sensitive method for metal detection; however, this must be supplemented with histologic and ultrastructural studies of the retrieved tissues to gain insight into the tissue localization and pathologic sequelae.

Particle Release and Distribution

Debris particles represent a substantial portion of metal degradation products generated by joint replacement prostheses. The degradation products of ceramics and polymers are exclusively in particulate form, as these classes of materials are generally considered insoluble in physiologic environments. Although polyethylene particles are generally recognized as the most prevalent particles in the periprosthetic milieu, metallic and ceramic particulate species are also present. When present in sufficient amounts, particles generated by wear, corrosion, or a combination of these processes induce the formation of an inflammatory, foreign-body granulation tissue with the ability to invade the bone-implant interface. This may result in progressive, periprosthetic bone loss that threatens the fixation of both cemented and cementless devices, limiting the survival of total joint replacements. Consequently, the role of debris particles of polymers, ceramics, and metal alloys in bone resorption and aseptic loosening has been the subject of intense study.

Less attention has been focused on particles generated by corrosion, perhaps because evidence of macroscopic corrosion in the current generation of single-part prostheses is rare. Recently, however, numerous reports have indicated that modular femoral total hip components can undergo severe corrosion at the tapered interface between the head and neck.[13-15] Several investigators have also observed that solid products of corrosion may form as a result of this process.[16] The clinical significance of corrosion at the modular head/neck junction lies, in part, in the effects that solid corrosion products may have at the bone-implant interface.

Although numerous investigations have focused on the presence of particulate debris in the periprosthetic tissues, relatively little is known about the dissemination of wear debris. Particles thought to be generated by a prosthetic device have been reported in distant organs of only a few patients with orthopaedic implants.[17] Implant-generated wear debris is difficult to identify in remote tissues, even in regional lymph nodes, because of the coexistence of particles from other sources.

In one recent study, wear particles that had disseminated beyond the periprosthetic tissue were primarily in the submicron size range.[18] Metallic wear particles were detected in the para-aortic lymph nodes in up to 70% of patients with total joint replacements. In some cases these particles further disseminated to the liver or spleen, where they were found within small aggregates of macrophages (Fig. 2) or as epithelioid granulomas throughout the organs. Metallic particles in the liver or spleen were more prevalent in subjects with previously failed arthroplasties (88% of such cases) than

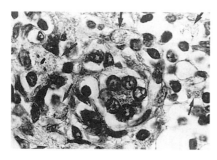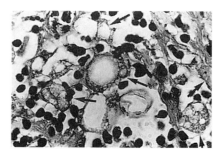

Figure 2 *Metal alloy particles within macrophages were found in the portal tracts of the liver of patients with total joint replacement components.* ***Left,*** *Light micrograph showing several pale staining, vacuolated macrophages (arrows) containing minute titanium particles as determined by electron microprobe analysis. The macrophages surround a bile duct in the liver of a patient with revision total hip arthroplasty. In this case, recurrent dislocation had resulted in severe wear of the titanium acetabular shell. The metallic particles were as large as 6 μm, but most were less than 1 μm in size (hematoxylin & eosin, × 1,250).* ***Right,*** *Light micrograph of macrophages (arrows) containing cobalt-chromium-molybdenum-alloy particles (as determined by electron microprobe analysis) in the portal tract of the liver of a patient who had hosted well-functioning, bilateral cemented total knee replacements for 15 years. Nearly all of the metallic particles were submicron in size (hematoxylin & eosin, × 1,250). (Reproduced with permission from Jacobs JJ, Goodman SB, Sumner DR, Hallab NJ: Biologic response to orthopaedic implants, in Buckwalter JA, Einhorn TA, Simon SR (eds): Orthopaedic Basic Science: Biology and Biomechanics of the Musculoskeletal System, ed 2. Rosemont, IL, American Academy of Orthopaedic Surgeons, 2000, pp 401-426.)*

patients with well-functioning primary joint replacements (average incidence of 18%).[18] Unlike polyethylene debris, metal particles can be characterized by electron microprobe analysis, which allows identification of individual particle composition against a background of particles from environmental or iatrogenic sources other than the prosthesis.

The smallest particles identified by the microprobe were approximately 0.1 μm in diameter. Dissemination of wear particles to liver, spleen, or abdominal lymph nodes is a common occurrence in patients with a total hip or knee replacement. Lymphatic transport is thought to be a major route for dissemination of wear debris. Wear particles may migrate via perivascular lymph channels as free or as phagocytosed particles within macrophages. Hematogenous dissemination of wear debris is also thought to occur. Particles may be transported to remote bone marrow by circulating monocytes or by entry of small particles directly into the bloodstream; these possibilities may be important in the migration of particles to end organs.

Hypersensitivity

Dermal hypersensitivity to metals is fairly common, affecting 10% to 15% of the population.[19] The term "hypersensitivity" refers to the induction of the immune system by a sensitizer. This response can be both humoral (initiated by antibody or formation of antibody-antigen complexes), which takes place within minutes (type I, II and III reactions) and cell-mediated (a delayed-type hypersensitivity response), which occurs over days (type IV reaction). Dermal contact and ingestion of metals have been widely documented to cause immune reactions.[19]

The incidence of metal sensitivity has been the subject of numerous investigations, albeit with heterogeneous patient populations and testing methodologies. The combined results of approximately 50 studies indicate that the prevalence of metal sensitivity in the general population is approximately 10% to 15%, with nickel sensitivity the highest (~14%) (Fig. 3).[19] Because the cross-reactivity of these antigens is high, the prevalence of metal sensitivity is generally considered to be 10%, the approximate average of the incidence of nickel, cobalt, and chromium sensitivity. Cross-reactivity between nickel and cobalt is the most common.

The incidence of metal sensitivity among patients with both well and poorly functioning implants is approximately twice as high (~25%) as that of the general population (Fig. 4). Furthermore, the prevalence of metal sensitivity among patients with a failed implant (based on five investigations) is 50% to 60% (Fig. 5), approximately five times the incidence of metal sensitivity found in the general population and two to three times that of all patients with metal implants.[19] While histologic studies of tissues recovered from failed total joint arthroplasties often report the presence of small numbers of lymphocytes, particularly in perivascular locations,[20,21] one recent study by Willert and associates[22] suggested that this lymphocytic response

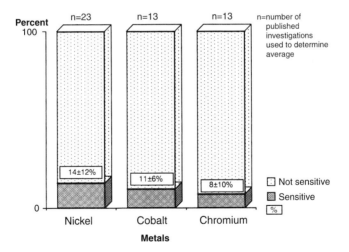

Figure 3 *Percentages of metal sensitivity among the general population for nickel, cobalt, and chromium. Data compiled from numerous published reports and listed as the average ± standard deviation. (Reproduced with permission from Jacobs JJ, Goodman SB, Sumner DR, Hallab NJ: Biologic response to orthopaedic implants, in Buckwalter JA, Einhorn TA, Simon SR (eds): Orthopaedic Basic Science: Biology and Biomechanics of the Musculoskeletal System, ed 2. Rosemont, IL, American Academy of Orthopaedic Surgeons, 2000, pp 401-426.)*

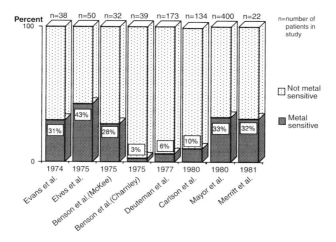

Figure 4 *A compilation of eight investigations in which patients with implants were tested for metal allergy by either patch or lymphocyte migration assay testing. Percentage is of patients with metal sensitivity after receiving a metal implant. (Reproduced with permission from Jacobs JJ, Goodman SB, Sumner DR, Hallab NJ: Biologic response to orthopaedic implants, in Buckwalter JA, Einhorn TA, Simon SR (eds): Orthopaedic Basic Science: Biology and Biomechanics of the Musculoskeletal System, ed 2. Rosemont, IL, American Academy of Orthopaedic Surgeons, 2000, pp 401-426.)*

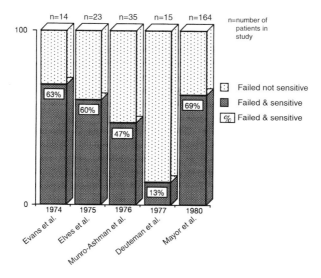

Figure 5 *A compilation of investigations showing the percentage of metal sensitivity among patient populations with failed or loose implants. (Reproduced with permission from Jacobs JJ, Goodman SB, Sumner DR, Hallab NJ: Biologic response to orthopaedic implants, in Buckwalter JA, Einhorn TA, Simon SR (eds): Orthopaedic Basic Science: Biology and Biomechanics of the Musculoskeletal System, ed 2. Rosemont, IL, American Academy of Orthopaedic Surgeons, 2000, pp 401-426.)*

may be more intense in failed, modern generation metal-on-metal devices and may be indicative of a cell-mediated immune reaction. This intense response was characterized by extended perivascular B and T lymphocytic infiltrations in secondary follicles with endothelial cell swelling, localized bleeding, and the expression of the proliferation-associated antigen Ki-67.[22] However, since perivascular lymphocytes are commonly found in failed metal-on-polyethylene implants,[20] as well as in patients with osteoarthritis prior to joint replacement (Thomas Bauer, MD, personal communication, 2000), the significance of these findings is unclear.

The increased prevalence of metal sensitivity among patients with loose prostheses has prompted speculation that immunologic processes may be a factor in implant loosening. At this time, however, it is unclear whether metal sensitivity causes implant loosening or implant loosening results in the development of metal sensitivity. Whether metal sensitivity is only an unusual complication in a few susceptible patients or a more common condition that plays a contributory role in implant failure has yet to be determined.

Carcinogenesis

The carcinogenic potential of the metallic elements used for joint replacement components is of particular interest because the large surface areas of cementless porous-coated devices are intended for implantation in younger, more active patients with life expectancies that may exceed 30 years. Animal studies have documented the carcinogenic potential of orthopaedic implant materials; small increases in rat sarcomas were found to correlate with metal implants that had high cobalt, chromium, or nickel content. Furthermore, lymphomas with bone involvement were more common in rats with metallic implants.[23] Implant site tumors—primarily osteosarcoma and fibrosarcoma—in dogs and cats have been associated with stainless steel internal fixation devices.[24]

The occurrence of tumors at the site of metallic implants has also been reported in humans. A review of the literature published before 1993 revealed 24 cases of malignancies adjacent to a total joint replacement. The most common lesion was malignant fibrous histiocytoma.[25] Four additional cases were reported in 1997.[26] Malignancies also have been reported adjacent to metallic internal fixation devices. A recent review listed only 11 reported cases, however.[27] Given the large number of joint replacement and internal fixation devices currently in situ, this is a relatively small number of cases, suggesting that the occurrence of peri-implant malignancies may be coincidental. Because many such cases may not be reported and these tumors may have relatively long latency periods, additional surveillance and broad-based epidemiologic studies are warranted.

Several human epidemiologic studies have examined systemic and remote cancer incidence in the first and second decades following total hip replacement. Two groups found slight increases in the risk of lymphoma and leukemia in patients with a cobalt-alloy total hip replacement, particularly in those patients who had a metal-on-metal device.[28,29] Larger, more recent series demonstrated no significant increase in leukemia or lymphoma;[30,31] however, these studies did not include a large proportion of subjects with metal-on-metal prostheses. Some investigators have actually documented a decreased incidence of certain tumors, including breast carcinoma, in recipients of total joint replacements.[29] This suggests that constitutive differences may exist between populations with and without implants that are independent of the implant itself, clearly confounding the interpretation of these epidemiologic investigations. At this point in time, the association of metal release from orthopaedic implants with carcinogenesis remains conjectural without any definite establishment of causality in humans. Because there is no relevant data on patient cohorts with mean follow-ups exceeding 15 years and the mean latency period for chromium-associated lung cancer following occupational metal exposure ranges from 13 to 30 years,[32] no definitive conclusions can be drawn from existing literature.

Relevance

When used for the appropriate indications and inserted with proper technique, orthopaedic implants have been quite successful; however, in certain situations adverse biologic effects may compromise the clinical outcome. Perhaps the clearest example is particle-induced periprosthetic bone loss leading to aseptic loosening—the most common complication of total joint arthroplasty. However, the toxicologic importance of remote and systemic findings is unknown. The clinical significance of orthopaedic wear debris accumulation at remote sites has been understood based largely on examination of lymph nodes biopsied at revision surgery or obtained for cancer staging in patients with a joint arthroplasty. Numerous case reports document the presence of metallic, ceramic, and polymeric wear debris from hip and knee prostheses in regional and pelvic lymph nodes along with findings of lymphadenopathy, gross pigmentation due to metallic debris, fibrosis, lymph node necrosis, and histiocytosis, including complete effacement of nodal architecture. The inflammatory response to metallic and polymeric debris in lymph nodes has included immune activation of macrophages and associated production of cytokines.[33]

In the liver and spleen, as in the lymph nodes, cells of the mononuclear-phagocyte system may accumulate small amounts of foreign materials without apparent clinical significance. However, heavy accumulation of exogenous particles can induce granulomatous lesions in the liver and spleen. The inflammatory reaction to particles in the liver, spleen, and lymph nodes is likely modulated as it is in other tissues, by material composition, size, the number of particles, their rate of accumulation, the duration of their presence, and the biologic reactivity of cells to these particles. Granulomatous lesions in the liver, spleen, and lymph nodes have been reported in response to heavy accumulation of metallic wear particles generated by joint replacements fabricated from cobalt or titanium alloys.[18] Accumulation of wear debris in the spleen and lymph nodes may explain, in part, recent observations suggesting that circulating peripheral blood monocytes from patients with joint replacements are more reactive to particulate wear debris stimulation than monocytes from individuals without implants.[34] Systemic monocyte hyperreactivity has been reported in other clinical settings such as the natural or surgical postmenopausal state.[35]

Future Directions for Research

Characterization of the bioavailability and bioreactivity of the metal species that have been released from metallic implants is the next logical step for investigation. Central to this determination is the speciation of the metal moieties present in body fluids and tissue stores that result from implant degradation, because many of the metals used in implants have valence and ligand-dependent toxicities in mammalian systems. Such studies represent an enormous challenge, given the technical complexities of working with concentrations in the parts per billion range. Current technological tools (graphite furnace Zeeman atomic absorption spectrophotometry and inductively coupled plasma-mass spectrometry) measure only the concentration of the element and provide no information on the chemical form or biologic activity. Current literature has very limited information describing the chemical form of the degradation products of metallic joint replacements. Ultimately, specific toxicologic investigation of relevant chemical species (as identified by bioavailability studies) should be used in animal models and cell cultures to delineate the biologic effects of these degradation products. Establishment of "toxicity thresholds" for these moieties in vitro and in vivo will enable the clinician to determine the toxicologic significance of elevated serum, urine, and/or remote tissue metal concentrations.

Metallic wear debris may extend into the nanometer size range,[36] suggesting that alternative methods of specimen preparation and analytic instrumentation may be required to more fully define the burden of metallic wear particles in remote tissues. Polyethylene particles also comprise a substantial portion of the disseminated wear particles in subjects with joint replacements. Although presence of larger (> 1 μm) polyethylene particles in the liver, spleen, and lymph nodes can be confirmed by Fourier Transform Infrared Spectroscopy microanalyses and/or microraman spectroscopy, smaller particles have so far eluded unequivocal identification. In these sites, differentiation from other birefringent endogenous and exogenous particles is impossible by polarized light microscopy alone. Advanced techniques for polyethylene detection in remote tissues are needed.

Finally, longer term multicenter epidemiologic studies are required to fully address the issues of metal implant-associated carcinogenesis, hypersensitivity, and remote toxicity. Before large scale studies are conducted, however, hypersensitivity testing modalities need to be standardized and correlated to local tissue response (both in the periprosthetic milieu and in reticuloendothelial storage sites) and clinical outcome. Future studies should also examine racial and gender differences in the systemic host response. Further advances in molecular biology and materials science, applied to the study of the host tissue response to implanted devices, may increase understanding of the critical determinants of implant biocompatability, providing new opportunities for the development of improved biomaterials, novel diagnostic and screening modalities, and pharmacologic strategies to modify host response. Ultimately, this may lead to improved clinical outcomes for patients requiring orthopaedic implants.

References

1. Williams DF: Biological effects of Titanium, in Williams DF (ed): *Systemic Aspects of Biocompatibility*. Boca Raton, FL, CRC Press, 1981, pp 169-177.

2. Gitelman HJ (ed): *Aluminum and Health, A Critical Review*. New York, NY, Dekker, 1989.

3. Jandhyala BS, Hom GJ: Minireview: Physiological and pharmacological properties of vanadium. *Life Sci* 1983;33:1325-1340.

4. Elinder CG, Friberg L: Cobalt, in Friberg L, Nordberg GF, Vouk VB (eds): *Handbook of the Toxicology of Metals: Specific Metals*, ed 2. Amsterdam, Netherlands, Elsevier, 1986, vol 2, pp 211-232.

5. Langard S, Norseth T: Chromium, in Friberg L, Nordberg GF, Vouk VB (eds): *Handbook of the Toxicology of Metals: Specific Metals*, ed 2. Amsterdam, Netherlands, Elsevier, 1986, vol 2, pp 185-210.

6. Sunderman FW: A pilgrimage into the archives of nickel toxicology. *Ann Clin Lab Sci* 1989;19:1-16.

7. Sunderman FW Jr, Hopfer SM, Swift T, et al: Cobalt, chromium, and nickel concentrations in body fluids of patients with porous-coated knee or hip prostheses. *J Orthop Res* 1989;7:307-315.

8. Michel R, Hofmann J, Löer F, Zilkens J: Trace element burdening of human tissues due to the corrosion of hip-joint prostheses made of cobalt-chromium alloys. *Arch Orthop Trauma Surg* 1984;103:85-95.

9. Jacobs JJ, Skipor AK, Patterson LM, et al: Metal release in patients who have had a primary total hip arthroplasty: A prospective, controlled, longitudinal study. *J Bone Joint Surg Am* 1998;80:1447-1458.

10. Jacobs JJ, Skipor AK, Black J, Urban RM, Galante JO: Release and excretion of metal in patients who have a total hip-replacement component made of titanium-base alloy. *J Bone Joint Surg Am* 1991;73:1475-1486.

11. Michel R, Nolte M, Reich M, Loer F: Systemic effects of implanted prostheses made of cobalt-chromium alloys. *Arch Orthop Trauma Surg* 1991;110:61-74.

12. Jacobs JJ, Skipor AK, Urban RM, et al: Systemic distribution of metal degradation products from titanium alloy total hip replacements: An autopsy study. *Trans Orthop Res Soc* 1994;19:838.

13. Mathiesen EB, Lindgren JU, Blomgren GGA, Reinholt FP: Corrosion of modular hip prostheses. *J Bone Joint Surg Br* 1995;73:345-350.

14. Collier JP, Surprenant VA, Jensen RE, Mayor MB, Surprenant HP: Corrosion between the components of modular femoral hip prostheses. *J Bone Joint Surg Br* 1992;74:511-517.

15. Gilbert JL, Buckley CA, Jacobs JJ: In vivo corosion of modular hip prosthesis in mixed and similar metal combinations: The effect of crevice, stress motion and alloy coupling. *J Biomed Res* 1993;27:1533-1544.

16. Urban RM, Jacobs JJ, Gilbert JL, Galante JO: Migration of corrosion products from modular hip prostheses: Particle microanalysis and histopathological findings. *J Bone Joint Surg Am* 1994;76:1345-1359.

17. Case CP, Langkamer VC, James C, et al: Widespread dissemination of metal debris from implants. *J Bone Joint Surg Br* 1994;76:701-712.

18. Urban RM, Jacobs JJ, Tomlinson MJ, Gavrilovic J, Black J, Peoc'h M: Dissemination of wear particles to the liver, spleen, and adominal lymph nodes of patients with hip or knee replacement. *J Bone Joint Surg Am* 2000;82:457-477.

19. Jacobs JJ, Goodman SB, Sumner DR, Hallab NJ: Biologic response to orthopaedic implants, in Buckwalter JA, Einhorn TA, Simon SR (eds): *Orthopaedic Basic Science: Biology and Biomechanics of the Musculoskeletal System*, ed 2. Rosemont, IL, American Academy of Orthopaedic Surgeons, 2000, pp 401-426.

20. Pizzoferrato A, Ciapetti G, Stea S, Toni A: Cellular events in the mechanisms of posthesis loosening. *Clinical Materials* 1991;7:51-81.

21. DiCarlo EF, Bullough PG: The biologic responses to orthopaedic implants and their wear debris. *Clinical Materials* 1992;9:235-260.

22. Willert H-G, Buchhorn GH, Fayyazi A, Lohmann CH: Histopathologische Veranderungen bei Metall/Metall-Gelenken geben Hinweise auf eine zellvermittelte Uberempfindlichkeit. *Osteologie* 2000;9:2-16.

23. Memoli VA, Urban, RM, Alroy J, Galante JO: Malignant neoplasms associated with orthopaedic implant materials in rats. *J Orthop Res* 1986;4:346-355.

24. Black J: *Orthopaedic Biomaterials in Research and Practice.* New York, NY, Churchill-Livingstone, 1988, pp 292-295.

25. Jacobs JJ, Rosenbaum DH, Hay RM, Gitelis S, Black J: Early sarcomatous degeneration near a cementless hip replacement: A case report and review. *J Bone Joint Surg Br* 1992;74:740-744.

26. Langkamer VG, Case CP, Collins C, et al: Tumors around implants. *J Arthroplasty* 1997;12:812-818.

28. Hinarejos P, Escuder MC, Monllau JC, et al: Fibrosarcoma at the site of a metallic fixation of the tibia: A case report and literature review. *Acta Orthop Scand* 2000;71:322-332.

29. Visuri T, Pukkala F, Paavolainen P, Pulkkinen P, Riska FB: Cancer risk after metal on metal and polyethylene on metal total hip arthroplasty. *Clin Orthop* 1996;329(suppl):S280-S289.

30. Gillespie WJ, Frampton CMA, Henderson RJ, Ryan PM: The incidence of cancer following total hip replacement. *J Bone Joint Surg Br* 1988;70:539-542.

31. Nyren O, McLaughlin JK, Gridley C, et al: Cancer risk after hip replacement with metal implants: A population-based cohort study in Sweden. *J Nat Cancer Inst* 1995;87:28-33.

32. Mathiesen EB, Ahlbom A, Bermann G, Lindgren JU: Total hip replacement and cancer: A cohort study. *J Bone Joint Surg Br* 1995;77:345-350.

33. Barceloux DG: Chromium. *Clinical Toxicology* 1999;37:173-194.

34. Hicks DC, Judkins AR, Sickel JZ, Rosier RN, Puzas JE, O'Keefe RJ: Granular histio-cytosis of pelvic lymph nodes following total hip arthroplasty: The presence of wear debris, cytokine production, and immunologically activated macrophages. *J Bone Joint Surg Am* 1996;78:482-496.

35. Lee S-H, Brennan FR, Jacobs JJ, Urban RM, Ragasa DR, Glant TT: Human mono-cyte/macrophage response to cobalt-chromium corrosion products and titanium parti-cles in patients with total joint replacements. *J Orthop Res* 1997;15:40-49.

36. Pacifici R, Vannice JL, Rifas L, Kimble RB: Monocytic secretion of interleukin-1 receptor antagonist in normal and osteoporotic women: Effects of menopause and estrogen/progesterone therapy. *J Clin Endocrinol Metab* 1993;77:1135-1141.

37. Doorn PF, Campbell PA, Worrall J, Benya PD, McKellop HA, Amstutz HC: Metal wear particle characterization from metal on metal total hip replacements: Transmission electron microscopy study of periprosthetic tissues and isolated parti-cles. *J Biomed Mater Res* 1998;42:103-111.

What guidelines/algorithms (both operative and nonoperative) are there for the treatment of osteolysis?

Osteolysis is the conclusion of a complex particle-induced biologic process resulting in bone loss and in some cases implant loosening.[1-5] Early diagnosis of osteolysis requires interval radiographic evaluation of patients with joint replacements. The success of early treatment underscores the need for consistent longitudinal surveillance.

The incidence of osteolysis is likely increasing as patients live longer and remain more active. Twenty-five years ago, a total hip replacement in a 65-year-old patient was expected to last for a lifetime. With increased longevity, however, hip replacements remain in service for many years. Significant activity demands can result in marked wear particle production.

Treatment of osteolysis in asymptomatic patients[6] is far different than treatment of patients with symptoms of a loose implant or pain from impending pathologic fracture. The goals of treatment of asymptomatic patients are preservation of bone stock, reduction of the risk of catastrophic periprosthetic fracture, and restoration of the articulation with the best material combinations currently available. Any treatment designed to address osteolysis must accomplish two key functions—debridement of the lesion, and modification of the articulation to decrease particle generation.[7] These treatment components are frequently referred to as the osseous lesion and the wear generator.

Although there is no clear consensus on the necessary frequency of radiographs following total hip replacement, most authors agree that images should be obtained 1 year after surgery and then at intervals of 1 to 2 years. When periprosthetic osteolysis is diagnosed, the radiographic evaluation should be more frequent in order to quantify the rate of progression and provide a clinical opportunity to question the patient for clues that might indicate a loose implant or impending pathologic fracture. The patient can also be informed about the process and its treatment. Patients are followed at 3- to 6-month intervals for several visits until a predictable pattern can be determined. If stable clinical and radiographic patterns are identified, then less frequent follow-up is acceptable. Each encounter should be used to improve patient understanding of the disease and its treatment. The addition of oblique pelvic views increases the sensitivity for the detection of osteolysis in total hip patients.

Acetabular Osteolysis

Several groups have proposed classifications of acetabular osteolysis. Paprosky and associates[8] developed a classification system for cemented implants that is based on the integrity of the acetabular rim and predicts the type of bone grafting that will be required to attain a stable implant. Type I defects involve minimal deformity. These are small, contained defects and are amenable to cancellous bone grafting. A cementless acetabular component, usually larger in diameter than the shell used in the primary arthroplasty, can be used to achieve stable fixation. Type II defects represent distortion of the normal acetabular hemisphere. In these defects, the anterior and posterior columns are intact although the medial wall and superior dome may be deficient. The reconstruction options include a high hip center, cancellous bone graft, femoral head allograft, and a variety of specialized components designed to replace deficient bone. Type III defects have severe bone loss requiring the use of structural allograft; examples include severe acetabular protrusio, column deficiencies, and pelvic discontinuity with associated global deficiency.

The American Academy of Orthopaedic Surgeons Committee on the Hip (COTH) classification has two basic categories, segmental and cavitary. Segmental defects result from complete bone loss in one area of the acetabulum and are subclassified into peripheral and central defects. Cavitary bone deficiencies represent contained bone loss with the acetabular rim remaining intact. This system has not been as useful as Paprosky's because it has not been combined with a specific treatment algorithm.

For uncemented acetabular components, a new classification system with treatment algorithm has been created (Fig. 1).[7] In the presence of acetabular osteolysis, type I implants are stable and have intact locking mechanisms. This combination can be treated with liner exchange, downsizing the femoral head if desirable, and cancellous bone grafting. Because lesions have been demonstrated to heal without grafting after liner exchange,[7] it is not always necessary to place cancellous graft into the lesions. Type II components are also stable; however, the function of the cup is compromised. Examples of Type II include a broken locking mechanism, backside wear of the shell so that the polyethylene liner would be unsupported or abraded, or component malposition. Treatment options include placement of a new polyethylene liner (possibly even a custom type) and acetabular shell retention with cementation of a new polyethylene liner. Acetabular shell removal may be necessary. Preoperative planning is critical if an existing component is to be left in place. Type III implants are loose and may have migrated. Acetabular screws should be removed at the time of revision to allow for stability testing and to permit improved access for lesion debridement. If grafting is to be performed, sufficient access may be achieved through the screw holes but may require a pelvic window. Debridement through the shell may be accomplished by curettage. Cancellous bone grafting or placement of bone graft substitutes is done through the same access. It is not yet known if grafting is required for some of these smaller lesions to heal.

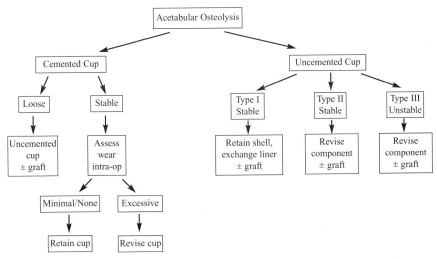

Figure 1 *Treatment of acetabular osteolysis. (Adapted with permission from Rubash HE, Sinha RK, Maloney WJ, Paprosky WG: Osteolysis: Surgical treatment. Instr Course Lect 1998;47:321-329.)*

Removal of the continuous source of particles is necessary. For cemented cups, the treatment depends on whether the implant is loose or stable (Fig. 1).

Femoral Osteolysis

The classification of femoral osteolysis proposed by the COTH has been useful. Femoral defects are classified as segmental, cavitary, or combined lytic. Segmental defects are characterized by erosion of the cortical bone, and are subclassified into complete or partial deficiencies. Cavitary defects represent contained lesions with destruction of endosteal bone. Combined defects, which include some element of both segmental and cavitary bone loss, are the most common type of femoral bone loss from osteolysis.

Treatment of asymptomatic femoral osteolysis varies with the extent and the progression. The extent of lytic changes should be carefully delineated; extensive radiographic evaluation may be required. Comparison of early postoperative radiographs with the most recent images demonstrating lysis is critical. This important comparison has become increasingly difficult, however, because of newer digital film storage and destruction of older radiographs. The economic demands of the health care environment must not lead to destruction of these important early evaluations.

The surgeon must determine if the femoral component is loose, if there is an impending pathologic fracture, and if there is eccentric polyethylene wear, all of which are relative indications for surgical intervention. In the presence of osteolysis, two special situations suggest more expedient surgical treatment–a loose or debonded femoral component with a rough surface texture, and a worn acetabular component with thin polyethylene.

Femoral osteolysis treatment goals are to maximize fixation and overall bone quality and minimize iatrogenic bone loss from revision surgery.

Treatment algorithms can be developed for femoral revision after considering the stability of the implant and its fixation, the need for cancellous and structural grafting, and the surgical approaches for removal (Fig. 2).

Pharmacologic Treatments

As our understanding of the biologic sequences that lead to osteolysis improves, nonsurgical treatment of osteolysis may become possible. Investigators have studied the use of nonsteroidal anti-inflammatory drugs and even gene transfer techniques to inhibit the inflammatory component of osteolysis.[9-11] In addition, newer medical therapies have been developed that may interrupt osteolytic progression. Shanbhag and associates[12-14] demonstrated in a canine model that bisphosphonates such as alendronate can be successfully used to prevent osteolysis. Although the peri-implant bone resorption was prevented in their model, the inflammatory response was not affected. This finding was not unexpected because bisphosphonates exert their effect primarily on the osteoclasts with no known anti-inflammatory effects. The efficacy of this drug in treating periprosthetic osteolysis is currently being evaluated in a multicenter placebo-controlled clinical trial. Such pharmaceutical therapy may be clinically useful in the early stages of osteolysis before the lesion has compromised implant stability, or in cases where surgery is either too complex or risky.

Relevance

Periprosthetic osteolysis is a progressive disease that requires careful radiographic evaluation. Progression of osteolysis to the level of substantial bone loss, impending pathologic fracture, or to the degree that future reconstruction would be compromised are indications for revision surgery. Limited revision surgery that involves exchange of the polyethylene liner, debridement of osteolytic lesions, retention of a stable femoral component, and

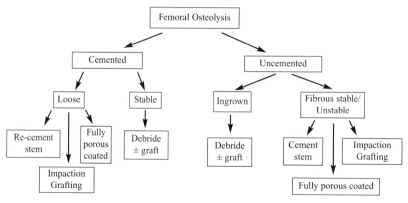

Figure 2 *Treatment of femoral osteolysis. (Adapted with permission from Rubash HE, Sinha RK, Maloney WJ, Paprosky W:, Osteolysis: Surgical treatment. Instr Course Lect 1998;47:321-329.)*

placement of a new femoral head have been successful in appropriate cases. The indications for bone grafting of acetabular and femoral lesions are not well defined. Finally, nonsurgical treatment of osteolysis could become a clinical reality.

Future Directions for Research

Will the newer materials (highly elevated cross-linked polyethylenes, ceramic-ceramic, and metal bearings) prevent osteolysis? If lysis does occur, will the clinical manifestations be similar? What modifications can be made to implant surfaces to prevent periprosthetic osteolysis? Is there an optimal articulation—lowest wear, least osteolysis, best range of motion and stability—and how should it be defined? Should advanced imaging techniques such as radiostereometric analysis or in-office digitized radiography be routinely used to detect early component wear? Are there low risk, inexpensive medical therapies for osteolysis?

References

1. Goldring SR, Schiller AL, Roelke M, Rourke CM, O'Neil DA, Harris WH: The synovial-like membrane at the bone-cement interface in loose total hip replacements and its proposed role in bone lysis. *J Bone Joint Surg Am* 1983;65:575-584.
2. Goodman SB, Chin RC, Chiou SS, Schurman DJ, Woolson ST, Masada MP: A clinical-pathologic-biochemical study of the membrane surrounding loosened and non-loosened total hip arthroplasties. *Clin Orthop* 1989;244:182-187.
3. Harris WH: The problem is osteolysis. *Clin Orthop* 1995;311:46-53.
4. Jiranek WA, Machado M, Jasty M, et al: Production of cytokines around loosened cemented acetabular components: Analysis with immunohistochemical techniques and in situ hybridization. *J Bone Joint Surg Am* 1993;75:863-879.
5. Schmalzried TP, Jasty M, Harris WH: Periprosthetic bone loss in total hip arthroplasty: Polyethylene wear debris and the concept of the effective joint space. *J Bone Joint Surg Am* 1992;74:849-863.
6. Maloney WJ, Herzwurm P, Paprosky W, Rubash HE, Engh CA: Treatment of pelvic osteolysis associated with a stable acetabular component inserted without cement as part of a total hip replacement. *J Bone Joint Surg Am* 1997;79:1628-1634.
7. Rubash HE, Sinha RK, Maloney WJ, Paprosky WG: Osteolysis: Surgical treatment. *Instr Course Lect* 1998;47:321-329.
8. Paprosky WG, Perona PG, Lawrence JM: Acetabular defect classification and surgical reconstruction in revision arthroplasty: A 6-year follow-up evaluation. *J Arthroplasty* 1994;9:33-44.
9. Goodman SB, Chin RC, Chiou SS, Lee JS: Suppression of prostaglandin E2 synthesis in the membrane surrounding particulate polymethylmethacrylate in the rabbit tibia. *Clin Orthop* 1991;271:300-304.
10. Spector M, Shortkroff S, Hsu HP, Lane N, Sledge CB, Thornhill TS: Tissue changes around loose prostheses: A canine model to investigate the effects of an antiinflammatory agent. *Clin Orthop* 1990;261:140-152.
11. Wooley PH, Sud S, Robbins PD, Whalen JD, Evans CH: Contrasting effects of gene therapy to inhibit interluken-1ß or tumor necrosis factor alpha in the murine inflammatory response to wear particles. *Trans Orthop Res Soc* 1998;23:122.

12. Shanbhag AS, Hasselman CT, Rubash HE: The John Charnley Award: Inhibition of wear debris mediated osteolysis in a canine total hip arthroplasty model. *Clin Orthop* 1997;344:33-43.

13. Shanbhag AS, Jacobs JJ, Black J, Galante JO, Glant TT: Macrophage/particle interactions: Effect of size, composition and surface area. *J Biomed Mater Res* 1994;28:81-90.

14. Shanbhag AS, Jacobs JJ, Glant TT, Gilbert JL, Black JM, Galante JO: Composition and morphology of wear debris in failed uncemented total hip replacement. *J Bone Joint Surg Br* 1994;76:60-67.

What are the best outcome measures for wear?

Outcome tools are used to determine the success of an intervention based on goals important to the individuals or populations being surveyed. Joint-specific evaluation tools differ in scope and objectives from those outcome measures designed to determine the effects of the intervention on a patient's lifestyle or a population's benefit from a particular intervention. It seems logical that wear mechanisms leading to failure are more likely to be detected by the joint-specific measures (eg, Harris Hip Score, Charnley hip score, Hospital for Special Surgery knee score, and Knee Society rating scores) than by the broader based outcome tools, such as the SF-36 or WOMAC outcome measures. To date, however, no outcome measures appear able to identify parameters related to wear and osteolysis—perhaps because of the insidious nature of the problem.

Wear represents only one of a number of mechanisms of failure of a total joint arthroplasty. Many factors influence the production of wear debris and the in vivo response of a particular patient. Similarly, multiple variables can influence failure of an arthroplasty. To evaluate wear accurately, measurement methods require sufficient sensitivity to identify wear as the major mode of failure and enough specificity to either eliminate or mitigate the influence of other factors contributing to failure of the arthroplasty. These tools will be of particular importance in the evaluation and development of new approaches to extend implant longevity. Because wear is an expected consequence of functioning mechanical devices, the surgeon should employ surgical and implant design approaches that minimize wear of devices and limit the consequences of wear debris.

Performance measurements of an implanted device include clinical tools (patient history and physical examination) and imaging tools, primarily radiographic evaluation to document the in vivo performance of the device and surrounding bone and soft tissues. There are as yet no predictable biologic measures of wear.

Clinical Tools

Although the patient history and physical examination remain key to identification of failing devices, these tools have major limitations. Wear debris generation, unless associated with catastrophic failure of the device, is asymptomatic. Clinical tools must be applied regularly, preferably with a joint-specific measuring tool. The history and physical examination may be sensitive to failure of the intervention, yet not be specific to identify wear as the source of that failure. Even if failure due to wear is identified, the joint may not be sufficiently painful or functionally limiting to warrant additional intervention. The clinical examination may not be sensitive enough to assess success or failure of devices solely based on wear. Finally, clinical symptoms and physical signs usually follow the diagnosis of wear-related failure by radiographic means; rarely would they precede the radiographic evidence.

Even with all these limitations, several clinical findings may suggest wear as the primary mechanism of failure. These findings are usually joint-specific, and may be identifiable in focused evaluations of the joint.

For the hip joint, clinical findings identifying wear as a failure mechanism include pain that is constant and consistent with debris-induced synovitis; start-up pain or catching, with patients requiring seconds to minutes before initiating ambulation from a seated position (indicating cup/liner instability); and a "clunking" sensation, suggesting uneven polyethylene surfaces. Inability to move the hip can indicate acute, catastrophic failure (liner displacement from shell, polyethylene fracture, or fracture of neck of the femoral component).

In the knee joint, clinical findings most suggestive of wear in a previously well-functioning knee replacement are unexplained effusion (culture negative) and instability. These findings are not specific because an insufficiently balanced or malaligned knee may initiate wear; both these findings may be present prior to the initiation of wear. Massive effusions are usually associated with more pronounced degrees of wear. Audible "grating" of components from metal-to-metal contact occurs with worn-through components or displacement of polyethylene components due to wear-associated loss of locking mechanisms.

Imaging Tools

Radiographic and digital analysis tools provide a greater degree of specificity for determining that wear is the mode of failure. Promising new techniques to assess in vivo wear before the development of radiographically obvious osteolysis or loosening are currently being evaluated.

Signs of early wear in the hip and knee are often difficult to detect on routine follow-up radiographs. Radiographic changes of less than 2 to 3 mm are hard to detect without the use of more sophisticated techniques, and require consistency in radiographic acquisition (positioning of patient, exposure of limb, magnification, etc). Standardization of radiographic techniques is a challenge in many practice environments. Most radiographic findings seen in common clinical practice reflect more advanced wear and are indirect. These findings reflect the reaction to the wear process rather than the wear itself. In the hip, osteolysis can occur in the proximal femur or more distally at mid- or distal portions of the stem. Acetabular osteolysis can occur at the margins of the implant, through screw holes or around screws, or extensively in the surrounding ilium, ischium, or pubis. Loosening of the acetabular component, with or without migration of the implant, can be a consequence of the wear process. Osteolysis in the knee can be seen at the margins of the tibial component, through or around screw holes, behind the femoral component with extension into the condyles around the patellar component, and can be seen with implant fracture (with corresponding loss of bone support) (Fig. 1). Other radiographic signs of polyethylene wear include progressive eccentric position of the head within the acetabular component, uneven wear of tibial or patella articular surfaces with recurrence of malalignment, or, in cases of catastrophic failure, displacement of parts (eg, a displaced patellar or acetabular component) (Fig. 2).

Early detection of wear at the bearing surfaces requires more sophisticated measurement techniques. The techniques available to assess wear radiographically require imaging approaches that can reproducibly measure changes in implant position of 0.1 to 0.3 mm. One major area of activity is computer-assisted assessment of plain radiographs to measure wear of acetabular components.[1-4] This approach represents a major improvement over templating techniques.[5] Edge-detection techniques with 2- and 3-dimensional analysis algorithms are used to measure penetration of the femoral head into the acetabulum, and are valuable tools in studying the performance of new developments in improving wear resistance. Edge-detection techniques require relatively straightforward positioning of patients and can accommodate technical variations in acquisition. At present, these tools are only available to evaluate conventional (metal-on-polyethylene) total hip arthroplasty; it is doubtful they can be useful in evaluating wear of the newer hard-hard (metal-to-metal or ceramic-to-ceramic) bearings.

Radiostereometric assessment (RSA) has become increasingly useful in analyses of both total hip and total knee implants. The primary use of RSA has been to observe movement of implants relative to bone in hip and knee replacements. Acquisition of information requires implantation of tantalum beads and standardized radiographic assessments with calibrated frames.[6-11] Several centers in the US are actively pursuing RSA; its usefulness in the

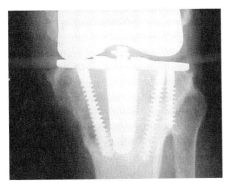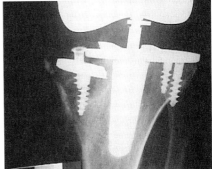

Figure 1 **Left,** *Primary total knee arthroplasty 5 years after implantation has medial joint-space narrowing and massive tibial osteolysis. The radiograph indicates advanced wear with debris-induced osteolysis.* **Right,** *Wear went unrecognized in this patient with a similar device. It resulted in massive bone loss, loss of implant support medially, and subsequent fracture of the tibial base plate.*

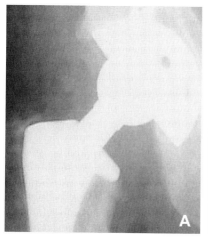
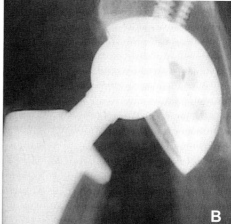
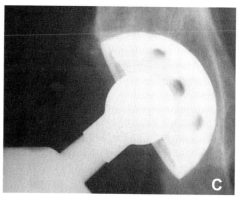

Figure 2 A, *Primary total hip arthroplasty 7 years after implantation in a young, heavy male patient with osteonecrosis. Eccentric position within the acetabulum demonstrates excessive wear.* **B,** *Four months later, the patient experienced acute pain, inability to ambulate, and radiographic evidence of a shift of the head without dislocation. The radiograph showed acute, catastrophic displacement of the polyethylene liner. The 32-mm head was titanium alloy.* **C,** *Three months after revision of the acetabular liner and the femoral head. The acetabular screws have been removed, and the new insert now articulates with a 28-mm CrCo femoral head.*

measurement of wear in total hip and total knee arthroplasty will be watched closely.

Several techniques have recently been suggested to study the indirect signs of wear, specifically peri-implant osteolysis. In the hip, these approaches have focused on positional modifications in the acquisition phase (oblique views) to more clearly identify bone loss behind the acetabular component. Spiral CT-based algorithms to assess osteolysis have also been developed.[12] These approaches may help quantify patient responses to wear debris. Such sophisticated measures are not presently available to study osteolysis around total knee arthroplasties.

Relevance

Wearing of components is an expected consequence of joint replacement. Few clinical and radiographic features seen in daily clinical practice will alert the orthopaedic surgeon to early failure due to wear. Most clinical and radiographic features become apparent in the more advanced stages and are usually detected by indirect radiographic signs of the effects of wear debris on the surrounding tissues. Careful analysis of new symptoms in a previously well functioning total joint arthroplasty and serial radiographic follow-up on a timely basis (every 12 to 24 months) should identify the large majority of impending failures. For new product development, stricter and more sophisticated evaluation should be required, using 2- and 3-dimensional radiographic techniques.

Future Directions for Research

Efforts to extend or develop validated outcome measures to address implant performance and wear should be continued. Relevant radiographic methodologies must be developed that can be applied in daily orthopaedic practice to identify the onset of wear-related component failure.[13] Increasingly sophisticated tools are needed to assess wear radiographically to aid new product development in both total hip and total knee arthroplasty. Quantitative methods with accuracy of 0.1 to 0.3 mm are required. Methods are also needed to assess wear in devices with alternative bearing surfaces.

References

1. Devane PA, Bourne RB, Rorabeck CH, Hardie RM, Horne JG: Measuring polyethylene wear in metal-backed acetabular cups: I. Three-dimensional technique. *Clin Orthop* 1995;319:303-316.
2. Devane PA, Bourne RB, Rorabeck CH, MacDonald S, Robinson EJ: Measuring polyethylene wear in metal-backed acetabular cups: II. Clinical application. *Clin Orthop* 1995;319:317-326.
3. Martell JM, Berdia S: Determination of polyethylene wear in total hip replacement with the use of digital radiographs. *J Bone Joint Surg Am* 1997;79:1635-1641.
4. Shaver SM, Brown TD, Hillis S, Callaghan JL: Digital edge-detection measurement of polyethylene wear after total hip arthroplasty. *J Bone Joint Surg Am* 1997;79:690-700.
5. Livermore J, Ilstrup D, Morrey B: Effect of femoral head size on wear of the polyethylene acetabular component. *J Bone Joint Surg Am* 1990;72:518-528.
6. Hildy MB, Yuan X, Ryd L: The stability of three different cementless tibial components: A randomized radiostereometric study in 48 knee arthroplasty patients. *Acta Orthop Scand* 1995;66:21-27.
7. Karrholm J, Borsser B, Lowenhielm G, Shorrason F: Does early micromotion of femoral stem prostheses matter? Four 7-year stereoradiographic follow-ups of 84 cemented prostheses. *J Bone Joint Surg Br* 1994;76:912-917.
8. Linder L: Implant stability, histology, RSA and wear: More critical questions are needed. A viewpoint. *Acta Orthop Scand* 1994;65:654-658.
9. Ohlin A, Selvik G: Socket wear assessment: A comparison of three different radiographic methods. *J Arthroplasty* 1993;8:427-431.
10. Ryd L: Micromotion in knee arthroplasty: A roentgen stereophotogrammetric analysis of tibial component fixation. *Acta Orthop Scand* 1986;220:1-80.

11. Toksvig-Larsen S, Ryd L, Lindstrand A: Early inducible displacement of tibial components in total knee prostheses inserted with and without cement. *J Bone Joint Surg Am* 1998;80:83-89.
12. Puri L, Wixson R, Stern S, Stulberg SD, Hendrix RW: The use of spiral computed tomography for the diagnosis of acetabular osteolysis after total hip arthroplasty. *J Bone Joint Surg 2000*; HAP Paul Award paper, ISTA 2000, Berlin.
13. Lewis G: Commentary: Design issues in clinical studies of the in vivo volumetric wear rate of polyethylene bearing components. *J Bone Joint Surg Am* 2000;82:281-288.

What is the outcome of the treatment of osteolysis?

Medical professionals do not actually treat osteolysis itself; rather, they treat the results of osteolysis. This question might be more aptly rephrased as "What are the outcomes of the treatment of the results of osteolysis?" Treatment of the adverse effects of osteolysis on total hip, total knee, total elbow, total shoulder, total ankle, metacarpophalangeal, and interphangeal arthroplasties is primarily bone loss management. Bone loss reaches clinical significance in three characteristic forms—loss of implant fixation, pathologic fractures, and impending pathologic fractures. Because periprosthetic osteolysis is a clinically silent disease, any symptoms associated with the results of periprosthetic osteolysis generally reflect either loss of implant fixation or a pathologic fracture.

No nonsurgical treatments of osteolysis have proved to be effective. Accordingly, this discussion of outcomes will first consider the indications for surgery in the management of the complications of periprosthetic osteolysis.

Although it is commonly stated that progressive osteolysis is an indication for surgery, all periprosthetic osteolysis is progressive; the rate of progression, however, varies widely. There are four indications for surgery for periprosthetic osteolysis: a pathologic fracture, an impending pathologic fracture, the presence of symptoms that are generally associated with a pathologic fracture or loosening of a component, and development of extensive osteolysis that would compromise future attempts at surgical reconstruction. An example of the last category would be severe lysis in the proximal femur of a patient who is functioning well and is completely asymptomatic if progression of the lysis would call for a circumferential proximal femoral allograft rather than a less complicated reconstruction.

The outcomes of treatment for osteolysis vary widely, and are compounded by wide variations in the location and extent of the lysis and the problems associated with the anatomy of the different joints. The goals of surgery are: (1) to replace the loose component; (2) fix a fracture; or (3) repair, bypass, or replace a bone deficiency. Included in the third category are such issues as the use of wedges or grafts in the tibia, and the use of circumferential proximal femoral allografts.

The efficacy of surgical treatment must be judged on the success in arresting the lysis, reattaching components, restoring bone stock, restoring function, and relieving pain. By most standards, the efficacy of arresting the lytic process surgically is quite satisfactory. The procedure entails removal of the periprosthetic membrane and reduction of the debris generated.

Undebrided granulomas within the pelvis, for example, can eventually mineralize and remain quiescent following revision of the acetabular component.

In a study of revision of cemented femoral components for periprosthetic osteolysis using new cemented femoral components, only 7% of the patients had radiographic evidence of periprosthetic osteolysis 8 years later.[1] These data do not support the hypothesis that osteolysis commonly recurs around cemented revision components in patients who developed femoral lysis around their primary cemented femoral prostheses.

The hemispherical cementless acetabular component, either fixed with screws or press-fit, has revolutionized acetabular revision surgery for lysis. The use of femoral fixation with porous-coated or rough-surfaced femoral implants, proximally porous-coated implants, rough-surfaced proximal and distal femoral implants, and improvements in cementing techniques have all played important roles in the success of revision surgery for lysis.

The widespread use of strut and cancellous allografts has improved the ability to restore bone stock. Impaction grafting is a specific example of the use of particulate cancellous allografts in total hip replacement surgery for the treatment of periprosthetic osteolysis. Bulk allografting around the acetabular component and the use of circumferential allografts around the hip and knee enable major restorations under extremely difficult circumstances. Nevertheless, these extensive procedures produce a high incidence of early and late complications. In addition, progressive resorption of bulk grafts has led to late failures.[2-4]

Relevance

The techniques described above have been effective in improving functional restoration and relief of pain. The use of extended trochanteric osteotomy and hooked trochanteric plates have also enhanced functional restoration. These new techniques have contributed substantially to a lower incidence of nonunion of the greater trochanter. Furthermore, modular components, offset liners, and components with varying offsets increase the available options at revision surgery. For the most severe cases, however, restoration of function may be incomplete and unsatisfactory.

Major revisions for severe bone loss involve long complex operations, consume substantial institutional resources, require high surgical skills, and produce uncertain outcomes. This type of surgery might best be done at specialized centers.

Directions for Future Research

The most important factor in preventing osteolysis is reduction of the burden of submicron wear debris. This requires better articular surfaces and the reduction of debris from other sources.

Longer and larger outcomes studies are needed, with particular emphasis on the prevalence of recurrent osteolysis and the functional limitations of more complex revisions. Autopsy retrievals following revision surgery and massive grafting are needed. Earlier revision operations are being recommended to avoid some of the severe problems, but the efficacy of this approach needs to be investigated.

There is an urgent need for better modes of revision surgery, such as adjuvant forms of increasing osteogenesis for the attachment of components to the skeleton, improved grafting techniques, use of bone substitutes, design improvements, and improvements in surgical technique. Tools are needed to evaluate the outcomes of all these new developments.

Grafting will clearly play an expanded role in the future, with increased use of bone substitutes such as calcium phosphate compounds, coral-type compounds, and the development of composite biologic matrices that incorporate the transduction signals.

References

1. Pierson JL, Harris WH: Cemented revision for femoral osteolysis in cemented arthroplasties. *J Bone Joint Surg Br* 1994;76:40-44.
2. Sloof TJ, Buma P, Schreurs BW, Schimmel JW, Huiskes R, Gardeniers J: Acetabular and femoral reconstruction with impacted graft and cement. *Clin Orthop* 1996;324:108-115.
3. Haddad FS, Spangehl MJ, Masri BA, Garbuz DS, Duncan CP: Circumferential allograft replacement of the proximal femur: A critical analysis. *Clin Orthop* 2000;371:98-107.
4. Shinar AA, Harris WH: Bulk structural autogenous grafts and allografts for reconstruction of the acetabulum in total hip arthroplasty: Sixteen-year average follow-up. *J Bone Joint Surg Am* 1997;79:159-168.

Biologic Issues

What are the local and systemic biologic reactions to wear debris?

What are the mediators (cellular, molecular, etc) of the local and systemic biologic reactions to wear debris?

Are there host factors that determine/modulate the biologic response to wear particles?

What specific features of wear particles are most important in determining the adverse biologic reactions?

What is the role of endotoxin and fluid pressure in osteolysis?

What experimental approaches (tissue retrieval, in vivo, in vitro, etc) have been used to investigate the biologic effects of particles?

Are there biologic markers of wear?

What potential biologic treatments are there for osteolysis?

What are the local and systemic biologic reactions to wear debris?

The issues concerning optimal materials and fixation techniques for total hip replacement (THR) are both complex and controversial. Charnley's concept of low friction arthroplasty was based on use of ultrahigh molecular weight polyethylene acetabular and stainless steel femoral components, which were attached to bone with the use of polymethylmethacrylate bone cement. This concept has been very successful; the large Scandinavian national hip replacement registers also indicate that this concept, with some modifications, gives very good 10-year results. However, aseptic loosening remains the singlemost important reason for failure of THRs. Recent data also indicate that low virulent bacteria, such as *Actinobacillus actinomycetemcomitans* and *Propionibacterium acnes*, can be found in some loose THRs.[1] Polymerase chain reaction techniques are likely to help delineate the role of infection.

Interface Versus Pseudosynovial Membrane

The adverse effect of mechanical loading in the loosening process has been emphasized; because most prosthetic components are rigid and their elasticity differs greatly from the host bone, some cases of aseptic loosening are likely to be purely mechanical.[2,3] The pumping of fluid around THRs leads to penetration of the pseudosynovial joint fluid along the implant interface. Research regarding biological loosening of the implants has primarily focused on the interface tissue.[4,5] However, the total mass of biologically active cells/tissues remains relatively small in the interface tissue compared to the pseudocapsule, where the lining layer is rich in monocyte/macrophage-like and fibroblast-like cells. The pseudosynovial joint fluid is thus a reservoir of biologically active agents important in the context of implant loosening, and extensive lysis may occur around mechanically well-fixed and clinically asymptomatic THRs.

Histologically, the capsular tissues show signs of organization into synovial-like structures with a lining cell layer of fibroblast-like type B and macrophage-like type A cells that grow directly, without a separating basement membrane, on loose connective tissue. This synovial cell layer produces hyaluronan, which is found in the pseudosynovial joint fluid in concentrations very near to those observed in normal diarthroidal joints. Monocyte/macrophages, fibroblasts, vascular endothelial cells, and lymphocytes also are present in these tissues. The local tissue reaction may result in aggressive bone lysis, which leads to a clinically, radiologically, and histologically distinct condition called aggressive granulomatosis.[6,7] More often, however, aseptic loosening is characterized by poorly vascularized connective tissue, where the main cells are fibroblasts. These cells have a different origin from cells of monocyte/macrophage lineage, which are bone marrow-derived.[8] Cytofluorographic studies have implied that cells synthesizing DNA are numerous in these tissues.[9]

The interface tissue in loosened implants is acidic and the pH values become very low in comparison to well-fixed cases. This has both physical and biologic relevance. Corrosion is accelerated and excess ionic wear is likely to occur, especially in the case of stainless steel. In chronically loose total hips, the pseudosynovial fluid is likely to become acidic. Due to the pumping effect generated by cyclic loading, this acidity comes into contact with the implant-bone interface, as well as the interface between the taper on the stem and the femoral ball.

Biomaterials and Design

In order to reduce wear and to achieve better biocompatibility, newer metallic alloys of cobalt chrome or titanium have been employed;[10] however, the theoretical benefits have yet to be converted into improved clinical outcomes. Low friction and wear are desirable characteristics of materials for articulating surfaces of THR prostheses. Metal-on-metal and metal-on-polyethylene are two classic designs that have, to a great extent, met the complicated tribologic challenge in THR. Major research efforts have focused on the optimization of tribology, elasticity, and biocompatibility of THRs by improvement in materials, macrogeometry, microtopography, and surgical techniques. With three decades of fundamental studies, the research techniques used to study the host response and mechanical wear have improved and include several useful innovations.[11] The bone harvest chamber has proved successful in studying local host response to wear products of implant materials in bone.[2] Radiostereometry reflects the quality of mechanical and biological implant fixation and can predict the outcome of THR at 2 years after implantation.[3] Hip simulators can provide useful preclinical data. However, the current materials for THR are far from optimal.

In the Finnish national arthroplasty register, roughly half of the total hip arthroplasties in recent years have been performed with cementless fixation. This has provided an excellent means to compare cemented and cementless fixation with respect to wear and loosening. Cemented prostheses still have better long-term outcomes. Bone cement is a viscoelastic material that may undergo creep; it is much stronger in compression than in tension, but weak in shear.[10] Cement in solid form is biocompatible, and the cement may become osseointegrated. In particulate form, bone cement becomes phagocytized and thus stimulates macrophage activation. Furthermore, the particles used as the contrast medium, such as barium sulfate or zirconium oxide, are abrasive and cause third-body wear in the articulating surface. Because of their size and shape, the contrast medium particles are potent stimulators for macrophage activation. Further consideration should be given to improvement of less abrasive contrast media.

Osteolysis was once thought to be an effect of cement particles, leading to the introduction of the term "cement disease." Cement wear particles are of varying size and are immunologically inert.[12] Periprosthetic osteolysis is not an entity specifically coupled to bone cement; "particulate" disease is a more appropriate term in this context. Corrosive and ionic wear particles have recently become a subject of interest, and electrical and pH-related processes in THR are now important research topics. The use of modified bone cement with lower heat generation has been unsuccessful and experience with Boneloc bone cement gave very high early failure rates.[10] While it has been very difficult to improve the qualities of current bone cements, educational efforts to improve cementing techniques have been associated with better long-term outcomes. Controversy still remains as to whether femoral components should be bonded to cements. Some advocate maximizing the bond between the surface of the stem and cement, while others advocate a polished surface with minimal bonding. Precoated stems have not uniformly met with success, and in some series polished stems did better than nonpolished ones.

Although polyethylene has several advantages, including low friction and good shock absorption, it has a high average wear rate of 0.1 to 0.2 mm/year (50 to 100 mm^3/year, corresponding to hundreds of millions of wear debris particles released into the surrounding tissues). The polyethylene particles generated are very small in size and function as potent stimulators of macrophages and other cells.[7,13]

Surgical-grade stainless steel is now rarely used in THRs. Cobalt-chromium alloys are stronger and more corrosion resistant than stainless steel. For joint implants, these alloys are usually composed of 30% to 60% cobalt, 20% to 30% chromium, 7% to 10% molybdenum, and various amounts of nickel. In bulk form, cobalt-chromium alloys are biocompatible but all the metallic components in the alloy are cytotoxic to some degree. Because of good abrasive wear resistance, the number of wear particles remains low. Titanium-based prostheses have become very popular in uncemented designs. Titanium in bulk form is quite biocompatible; however, it is soft and wears easily. Thus the number of wear particles is very high and clinical metallosis develops in the surrounding tissues very quickly. In the articulating surfaces, titanium should not be used because the current commercial coating techniques have failed to reduce the wear and metallosis and excessive wear of polyethylene. In cementless designs, hydroxyapatite coatings seem to improve fixation, but they can contribute to abrasive third-body wear. Corrosion of the taper when cobalt-chromium or ceramic heads are used is a further drawback in titanium-based THRs.

In addition to wear debris from articulating surfaces, metallic materials dissolve in ionic or particulate form in the corrosive environment of body fluids.[14] Local high acidity is an important factor in the biologic wear process. It has been estimated that an implanted cobalt-chromium component dissolves at approximately 50 nm/year. Despite these shortcomings, these traditional metals, methylmethacrylate, and polyethylene will form the basis of THR prosthesis manufacturing technology in the near future.

Recently, amorphous diamond coatings have been used in total joint prostheses.[14,15] Diamond against highly cross-linked polyethylene is a promising articulation because the friction and wear are extremely low. Wear simulator tests with a diamond-on-diamond articulating system showed less than 10 nm wear after 5 million cycles. Such total hip prostheses generate few wear particles and potentially reduce the biological load. Diamond coatings can be applied to previously clinically tested implants without loss of their basic properties.

Macrophage Activation and Osteolysis

Once the particles are phagocytosed, the mono- and multinuclear phagocytic cells are activated. The function of these cells after phagocytosis is more aggressive than that of a resting cell[8] and is modulated by various soluble paracrine and autocrine stimuli. Furthermore, there are significant extracellular matrix-cell interactions and direct cell-to-cell contacts,[16] which affect the phenotype and function of these cells. The process involves more than recruitment and accumulation of inflammatory cells; the inflammatory environment may also downregulate cells that would be necessary to counteract the excessive foreign body reaction and mediate normal tissue remodeling. Such cells include neurones (or the peripheral extensions, axons) and endothelial cells. A second consequence of the multiple regulatory and signaling systems is that the activated phenotype of the phagocytic cells varies in different sites and at different times. Phagocytosis and activation of monocyte/macrophages lead to recruitment of many resident and immigrant cells in such a manner that peri-implant bone is destroyed. Considering the huge particle load, with over 10^5 particles of polyethylene being formed at each step, it is unclear why some patients develop osteolysis while others do not. Pertinent variables may include genetic predisposition; the number, size, charge, chemical composition, and location of the particles; and micromotion and shear forces from fluid (pseudosynovial fluid, bone extracellular fluid, etc) flux.

Impact of Cytokine Network

In loosening, periprosthetic bone support is lost. Those cytokines that regulate the phenotype and function of bone cells have therefore shifted so that the overall net function of the osteolytic osteoclasts overcomes that of the bone forming osteoblasts.[17] Cytokines seem to work as a network.[18] Some cytokines relevant to osteolysis are tabulated in Table 1. Information from substraction libraries and microarray techniques/DNA microchips may extend this list. Some cytokines probably play a more important role than others. Tumor necrosis factor-α (TNF-α) and interleukin-1 (IL-1α and IL-1β), for example, are considered to be particularly important for bone destruction. Modulation of TNF-α with chimeric, humanized TNF-α antibodies or with recombinant soluble p55 or p75 TNF receptors inhibits formation of bone erosions in rheumatoid arthritis. From the biologic point of

Table 1 Effects of cytokines on bone

Cytokines	Cells of origin	Target cells	Effects on bone
IL-1	Osteoblast Macrophage Fibroblast Lymphocyte Mast cell Endothelial cell	Osteoblast Osteoclast	Formation +/- Resorption +/-
IL-3	Macrophage Lymphocyte Mast cell	Osteoclast	Resorption -
IL-6	Osteoblast Macrophage Fibroblast Lymphocyte Mast cell Endothelial cell	Osteoblast Osteoclast	Formation - Resorption +
IL-8	Macrophage Fibroblast Lymphocyte Endothelial cell	Osteoblast Osteoclast	Resorption +
IL-11	Osteoblast Fibroblast Mesenchymal cell lineage	Osteoclast	Formation - Resorption +
TNF-α	Osteoblast Macrophage Fibroblast Lymphocyte Mast cell Endothelial cell	Osteoblast Osteoclast	Resorption +
TNF-β	Macrophage Lympocyte Mast cell	Osteoblast Osteoclast	Resorption +
M-CSF	Osteoblast Macrophage Fibroblast Lymphocyte Mast cell Endothelial cell	Osteoclast	Resorption +
GM-CSF	Osteoblast Macrophage Fibroblast Lymphocyte Mast cell Endothelial cell	Osteoblast Osteoclast	Resorption +

Continued

Table 1 Effects of cytokines on bone

Cytokines	Cells of origin	Target cells	Effects on bone
EGF	Macrophage Fibroblast Endothelial cell	Osteoblast Osteoclast	Formation + Resorption +
TGF-α	Macrophage Fibroblast Mast cell Endothelial cell	Osteoblast Osteoclast	Formation - Resorption +
TGF-β	Osteoblast Macrophage Fibroblast Lymphocyte Mast cell Endothelial cell	Osteoblast Osteoclast	Formation + Resorption +/-
PDGF	Osteoblas Macrophage Fibroblast Mast cell Endothelial cell	Osteoblast Osteoclast	Formation + Resorption +
aFGF	Osteoblast Macrophage Fibroblast Endothelial cell Lymphocyte	Osteoblast	Formation +
bFGF	Osteoblast Macrophage Fibrobast Endothelial cell	Osteoblast	Formation +
IGF	Osteoblast Macrophage Fibroblast Endothelial cell	Osteoblast Osteoclast	Formation + Resorption +
IFN-γ	Macrophage Fibroblast Lymphocyte	Osteoblast Osteoclast	Resorption -
BMP	Osteoblast Fibroblast	Osteoblast Osteoclast	Formation + Resorption -
RANKL	Osteoblast Fibroblast	Osteoclast	Resorption +
OPG	Osteoblast	Osteoclast	Resorption -

IL = interleukin, TNF = tumor necrosis factor, CSF = colony stimulating factor, M= macrophage, GM = granulocyte-macrophage, EGF = epidermal growth factor, TGF = transforming growth factor, PDGF = platelet-derived growth factor, FGF = fibroblast growth factor, IGF = insulin-like growth factor, IFN = interferon, BMP = bone morphogenetic protein, RANKL = receptor activator kappa B ligand (= osteoclast differentiation factor or ODF), OPG = osteoprotegerin.

+ = Stimulation: +/- = Stimulation or inhibition; - = Inhibition.

view, IL-1 may be even more important for the formation and function of osteoclasts. Studies with transgenic and knock-out animal models and with IL-1 receptors (IL-1RI and IL-1RII) or interleukin-1 receptor antagonists suggest that modulation of IL-1 might be useful in the prevention of pathologic bone destruction. These biologic agents currently are not used for the prevention and treatment of implant loosening; there are not yet enough clinical data and the treatment, which usually extends over many years, would involve considerable cost.

Although the practical applications seem distant, the most exciting new entries on the list deserve special mention, namely receptor activator of NF-κB ligand (RANKL, also known as osteoclast differentiation factor) and osteoprotegerin (OPG). RANKL can bind to its receptor (RANK) and promotes differentiation of monocytes to osteoclasts. Soluble OPG also binds RANKL and neutralizes its effect, thus inhibiting osteoclast formation. It is possible to control osteoclast formation and osteolysis with recombinant RANKL and OPG. Osteoclast formation has been induced in vitro with the combined use of macrophage colony stimulating factor (M-CSF) and RANKL in human peripheral blood monocyte cultures. Pseudosynovial fluid from patients with loosened THR implants appears to drive the same process. It is clear, however, that phagocytosis stimulates the formation of many important regulatory cytokines. These cytokines can stimulate osteolysis by osteoclasts and/or inhibit bone formation by osteoblasts at the level of precursor cell recruitment, or affect proliferation, differentiation, or activation.

Proteinases and Periprosthetic Tissue Destruction

Neutral endoproteinases of the matrix metalloproteinase (MMP) family have attracted considerable attention (Table 2).[19] Although not able to degrade mineralized bone, they may play an important role in the formation and progression of the synovial-like interface membrane. MMPs influence the migration of cells from blood vessels to tissues. Because of their ability to produce chemotactic peptides and to solubilize various cell membrane-associated molecules, MMPs may play a role in processing molecules such as TNF-α and RANKL. They may also participate in the activation-resorption-formation cycles after the osteoclast leaves the resorption pit and moves to a new site, leaving a demineralized surface behind; this surface has to be processed before osteoblasts attach to it. Phagocytosis and the subsequent cellular activation are associated with increased local production of proMMPs.

Another exciting new proteinase involved in periprosthetic bone lysis is an acidic cysteine endoproteinase, cathepsin K. The osteoclast is able to attach to bone with a tight junction. Osteoclasts express at αVβ3, αVβ5, α2β1, and αVβ1 integrin receptors, which may participate in adhesion, migration, and endocytosis. Cadherins may play a role in the tight attachment of osteoclasts to bone. Osteoclasts are able to produce an acidic sub-osteoclastic (potential) space, a Howship's lacuna. At the low pH prevailing

Table 2 Human matrix metalloproteinases (MMPs)

Name	MMP group	Other names
Collagenase-1	MMP-1	Interstitial collagenase, fibroblast-type collagenase
Collagenase-2	MMP-8	Neutrophil-type collagenase
Collagenase-3	MMP-13	
Gelatinase A	MMP-2	72 kD type IV collagenase
Gelatinase B	MMP-9	92kD type IV collagenase
Stromelysin-1	MMP-3	Transin, proteoglycanase
Stromelysin-2	MMP-10	
Stromelysin-3	MMP-11	
MT1-MMP	MMP-14	
MT2-MMP	MMP-15	
MT3-MMP	MMP-16	
MT4-MMP	MMP-17	
MT5-MMP	MMP-24	
MT6-MMP	MMP-25	Leukolysin
Matrilysin	MMP-7	PUMP-1
Matrilysin-2	MMP-26	Endometase
Metalloelastase	MMP-12	Macrophage elastase
Enamelysin	MMP-20	
RASI-1	MMP-19	
CA-MMP	MMP-23	

in the resorption lacuna below the ruffled border, bone hydroxyapatite $Ca_{10}(PO_4)_6(OH)_2$ is dissolved. This is followed by collagenolysis, which was earlier thought to be mediated by cathepsins like B, L, and H, but is now believed to be cathepsin K-driven. Cathepsin K is able to effectively cleave interstitial collagens and is not restricted to the [775]Gly-[776] Ile(Leu) site as are all the classical mammalian collagenases. Instead, cathepsin K has a bacterial collagenase-like action and is able to cleave effectively at multiple sites across the collagen triple helix.

Interestingly, the interface tissue loaded with foreign bodies also contains many mononuclear, CD68+, and TRAP-positive cells. The pH prevailing in the interface tissue is so low that it can demineralize the bone surface in contact with the acidic interface. This pH varies from around 5.5 to 6.5, which is also an optimum for the collagenolytic function of cathepsin K. This challenges the notion that lysis of bone type I collagen occurs mainly in the subosteoclastic space. Osteolysis might also occur at the contact surface between the acidic interface and host bone. Many of the mononuclear cathepsin K-positive cells contain phagocytosed debris. Foreign body-type giant cells are also cathepsin K-positive. Therefore, particle-stimulated activated monocyte/macrophages may directly participate in bone lysis around the implant.

Foreign implant-derived bodies participate in many other important pathobiologic processes relevant to loosening. Some of the effects are indirectly mediated through altered extracellular matrix-cell signaling and direct cell-to-cell contacts.[16,20]

Relevance

The primary local reaction results from the activation of monocyte/macrophages as a consequence of phagocytosis. This leads to local production of cytokines, which recruit both circulating and local resident cells into the interface membrane between the implant and bone, causing a paracrine activation of periprosthetic osteoclasts. Osteolysis leads to loss of periprosthetic bone support that, together with cyclic loading, finally leads to aseptic loosening of the implant. Toxic, carcinogenic, and teratogenic reactions generally do not occur with the biomaterials currently in use. However, sensitization and hypersensitivity reactions to implant-derived materials are possible. These may accelerate prosthetic loosening and also cause systemic symptoms. Thus, local and systemic biologic reactions can contribute to periprosthetic bone loss, pain, and hypersensitivity.

Future Directions for Research

Preclinical and clinical research is needed on ceramic-to-ceramic and ceramic-to-metal combinations, diamond surface coating, and newer cross-linked polyethylenes. These material strategies may diminish the particle load and inhibit the adverse host response ("particulate" disease).

Currently, little is known about the effects of particle phagocytosis on macrophages to enable pharmacologic strategies. Inhibition of excessive osteoclast function in periprosthetic bone is an attractive strategy, although it is an effector mechanism that is downstream from the initiating events. Experiments are ongoing with drugs such as bisphosphonates, tetracyclines, and their chemically modified tetracycline derivatives; cathepsin K inhibitors are being developed. Because the process is localized, it seems appropriate to develop some vehicles or methods to ensure a controlled localized release of these type of pharmacologic agents.

References

1. Glenn JV, O'Hagan S, Clingen-Vance G, Nixon J, Patrick S: Serodiagnosis in anaerobic periprosthetic hip infection. *Anaerobe 2000: An International Congress of the Confederation of Anaerobic Societies.* Manchester, England, 2000, p 206.
2. Goodman SB: The effects of micromotion and particulate materials on tissue differentation: Bone chamber studies in rabbits. *Acta Orthop Scand* 1994;258(suppl): 1-43.
3. Kärrholm J, Malchau H, Snorrason F, Herberts P: Micromotion of femoral stems in total hip arthroplasty: A randomised study of cemented, hydroxyapatite-coated and porous-coated stem with roentgen stereophotometric analysis. *J Bone Joint Surg Am* 1994;76:1692-1705.
4. Goldring SR, Clark CR, Wright TM: The problem in total joint arthroplasty: Aseptic loosening. *J Bone Joint Surg Am* 1993;75:799-801.
5. Harris WH: The problem of osteolysis. *Clin Orthop* 1995;311:46-53.

6. Santavirta S, Konttinen YT, Bergroth V, Eskola A, Tallroth K, Lindholm TS: Aggressive granulomatous lesions associated with hip arthroplasty. Immunopathological studies. *J Bone Joint Surg Am* 1990;72:252-258.

7. Santavirta S, Hoikka V, Eskola A, Konttinen YT, Paavilainen T, Tallroth L: Aggressive granulomatous lesions in cementless total hip arthroplasty. *J Bone Joint Surg Br* 1990;72:980-984.

8. Lassus J, Salo J, Jiranek WA, et al: Macrophage activation results in bone resorption. *Clin Orthop* 1998;352:7-15.

9. Santavirta S, Hietanen J, Ceponis A, Sorsa T, Kontio R, Konttinen YT: Activation of periprosthetic connective tissue in aseptic loosening of total hip replacements. *Clin Orthop* 1998;352:16-24.

10. Santavirta S, Konttinen YT, Lappalainen R, et al: Materials in total joint replacement. *Current Orthop* 1998;12:51-57.

11. Santavirta S, Takagi M, Gomez-Barrena E, et al: Studies of host response to orthopaedic implants and biomaterials. *J Long-Term Effects Med Implants* 1999;9:67-76.

12. Santavirta S, Konttinen YT, Bergroth V, Grönblad M: Lack of immune response to methylmethacrylate in lymphocyte cultures. *Acta Orthop Scand* 1991;62:29-32.

13. Santavirta S, Nordström D, Metsärinne K, Konttinen YT: Biocompatibility of polyethylene and host response to loosening of cementless total hip replacement. *Clin Orthop* 1993;297:100-110.

14. Santavirta SS, Lappalainen R, Pekko P, Anttila A, Konttinen YT: The counterface, surface smoothness, tolerances, and coatings in total joint prostheses. *Clin Orthop* 1999;369:92-102.

15. Nordsletten L, Hogåsen AKM, Konttinen YT, Santavirta S, Aspenberg P, Aasen AO: Human monocytes stimulation by particles of hydroxyapatite, silicon carbide and diamond: In vitro studies of new prosthesis coatings. *Biomaterials* 1996;17:1521-1527.

16. Konttinen YT, Li TF, Hukkanen M, Ma J, Xu JW, Virtanen I: Signals targeting the synovial fibroblast in arthritis. *Arthritis Res* 2000;2:348-355.

17. Konttinen YT, Waris V, Xu JW, et al: Transforming growth factor-beta 1 and 2 in the synovial-like membrane between implant and bone in loosening total hip artroplasty. *J Rheumatol* 1997;24:694-701.

18. Konttinen YT, Xu J-W, Pätiälä H, et al: Cytokines in aseptic loosening of total hip replacement. *Current Orthop* 1997;11:40-47.

19. Takagi M: Neutral proteinases and their inhibitors in the loosening of total hip prosthesis. *Acta Orthop Scand* 1996;67(suppl 271):1-29.

20. Li TF: Thesis: Extracellular and pericellular matrix proteins in the synovial membrane-like interface tissue from aseptic loosening of total hip replacement. Hensinki, Finland, University of Helsinki, 2000.

What are the mediators (cellular, molecular, etc) of the local and systemic biologic reactions to wear debris?

The importance of biologic mediators in aseptic loosening following total joint arthroplasty is evident in part from the speed and severity of localized bone resorption at the margins of previously well-fixed prostheses. Localized bone loss is generally associated with scalloping lesions of reduced bone mineral density that are apparent within 1 to 3 years after primary total joint arthroplasty. The focal loss of bone, often detected at multiple sites along the length of the prosthesis and during revision arthroplasty, accompanies a prominent foreign-body macrophage response to implant-derived polymeric and metallic wear particles. The onset of bone resorption may occur at vastly different rates among patients with the same type of implant. Variability in patient response to wear debris likely reflects a multifactorial process, which includes elements of genetically based cellular reactivity.

Wear Debris and Periprosthetic Osteolysis

Generation of wear debris occurs immediately after implant insertion and ultimately results in a profile of particles, typically less than 7 to 10 microns, that includes all total joint arthroplasty biomaterials.[1-4] The particles represent ultra-high molecular weight polyethylene, polymethylmethacrylate (PMMA) bone cement, pure elemental metals, and metal alloys.[5-8] The origin of the wear particles can be traced to both articulating and nonarticulating surfaces of the prosthetic materials.[9-13] The causes of particle generation vary from micromotion to more passive reactions such as corrosion or oxidative surface reactions.[14-16]

The extent of bone resorption at the implant-bone interface varies with the severity of a granulomatous tissue response to wear debris and determines the time lapse before implant loosening occurs.[17,18] Immunohistochemical studies of the inflammatory tissue from failed total joints show that macrophages are the predominant cell type in tissue from regions of extensive osteolysis.[19,20] The implant interface tissue also contains fibroblasts, lymphocytes, and other inflammatory cells.[19-21] Jiranek and associates[21] demonstrated that macrophages and fibroblasts are present in tissue at the cement-bone interface of loose cemented acetabular components. Polyethylene and PMMA wear particles were universally present in tissues from failed cemented acetabular components. Santavirta and associates[22] compared the tissue obtained from aggressive granulomatous lesions (osteolysis) to that obtained from cases of aseptic loosening without significant osteolysis. In the granulomatous lesions, the histology was characterized by a well-organized connective tissue containing macrophages and fibroblastic reactive zones. They proposed that the biologic response to particulate wear debris was a continuum of cellular responsiveness in which normal processes involving clearance and repair break down with high particle load. When these conditions are met, the chronic inflammatory processes cannot be "walled off", and focal osteolysis results.

Numerous in vivo and in vitro studies have shown that wear particle-induced macrophage activation plays a role in periprosthetic osteolysis.[1] Essentially, this occurs by two biologic mechanisms. First, wear particle-associated macrophages release pro-inflammatory factors (eg, cytokines, growth factors, prostaglandins) that enhance the activity of osteoclasts, the cells which carry out bone resorption. Second, osteoclasts are formed from mononuclear phagocyte precursors that are present in the wear particle-induced macrophage infiltrate. These processes are not mutually exclusive; other stromal and inflammatory cell elements found at the bone-implant interface likely influence both the extent of osteoclast formation and bone resorption.

Proinflammatory Mediators and Particle-Induced Osteolysis

The contribution of the cells present within the macrophage-rich inflammatory tissue to the induction of bone resorption and implant loosening involves multiple cellular mechanisms. Goldring and associates[23,24] demonstrated that tissue samples collected from failed implants release high levels of prostaglandin E2 (PGE2) and bone resorbing activity into the culture medium. The bone resorbing activity was only partially blocked by abolishing prostaglandin production by the cultured membrane with indomethacin, suggesting that multiple factors contribute to bone resorption. Kim and colleagues[25,26] showed that cultured membrane specimens from cementless and cemented prostheses release collagenase, gelatinase, PGE2 and interleukin (IL-1) into the medium. Membranes retrieved from hip prostheses inserted without cement showed that polyethylene articulations produce higher levels of collagenase and IL-1. IL-1α release was highest in membranes associated with radiographic evidence of focal osteolysis (endosteal erosion). Monocytes isolated from interfacial membranes demonstrated a "sensitized" reactivity to metallic particles when compared to cells from individuals without an implant.[27] Taken together, these data confirm the roles of multiple soluble mediators in the process by which bone resorption is sustained at the bone-implant interface.

Macrophage Response to Wear Debris

Cell culture studies have enabled quantification of the effects of specific types of particles on cell metabolism. Rae[28] demonstrated that cobalt particles are toxic to murine macrophages, while titanium, aluminum, and chromium particles are well tolerated. Maloney and associates[29] demonstrated that bovine synovial fibroblasts in primary monolayer cultures react differentially to similar sized particles of titanium, titanium-aluminum alloy, and chromium. Herman and colleagues[30] showed that peripheral blood mononuclear cells exposed to either bone cement particles or to a surface coating release IL-1, tumor necrosis factor (TNF), and PGE2 and that this results in increased release of ^{45}Ca from bone samples. Glant and Jacobs[31] analyzed the macrophage response to different sizes, concentrations, and composition of particles and showed that commercially pure titanium, PMMA, and polystyrene stimulate IL-1 and PGE2 release in dose-dependent and time-dependent manners. Titanium particles of phagocytosable size (1 to 3 μm) were most efficent in stimulating bone-resorbing activity at concentrations that corresponded to maximal release of IL-1. Anti-IL-1 antibodies and indomethacin only partly suppressed the bone resorbing activity of the macrophage-conditioned medium. Bone resorption in response to the conditioned medium of the macrophage cultures varied among different macrophage populations.[32] Studies of effects of PMMA particle size, type, and surface area on cellular reactivity showed that phagocytosable particles stimulate an increase in TNF release by the macrophages 12 times that of larger nonphagocytosable particles.[33]

Macrophage-Dependent Biologic Mediators

Macrophages migrate to sites of localized debris.[34-37] The macrophages are activated by the particles and subsequently release proinflammatory cytokines and other agents that induce bone resorption.[38-41] Macrophage products capable of inducing bone resorption include IL-1α, IL-1β, IL-6 and TNF-α, arachidonic acid metabolites, and degradative enzymes.[42-45] The existence of multiple factors at one site is likely to accelerate bone destruction.[46-48] Interleukin-1α and -1β and TNF-α may also induce secondary effects on the other cell types (such as osteoblasts) in the interfacial membrane,[49-53] resulting in release of matrix-degrading enzymes, including collagenase, stromelysin, gelatinases, and plasminogen activators. The release of the interleukins and TNF-α occurs in both macrophages in culture exposed to particulate debris and in retrieved interface membranes maintained in organ culture.[54-56]

The proinflammatory mediators associated with the periprosthetic tissue response represent a constellation of cytokines and other soluble factors. Synovial fluids associated with the periprosthetic tissue reaction reveal that PGE2, IL-1, and collagenase levels are elevated when compared to control fluid samples.[57,58] The soluble factors released include macrophage colony stimulating factor (CSF) and IL-6, both proteins that stimulate osteoclast differentiation.[59,60] Other factors such as transforming growth factor-β1 and -β2 are elevated in the membranes where their profibrotic and immunomodulatory potential may contribute to the persistence and expansion of the granulatomatous tissue.[61] Increased levels of platelet-derived growth factor A and B are also present in the macrophages and fibroblasts of the interfacial membrane and may influence adjacent bone cell metabolism.[62] Granulocyte/macrophage CSF has also been implicated in the cellular proliferation in the interfacial membrane around implants.[63] Other cytokines that may exhibit immunomodulatory roles include IL-12, which is increased in the pseudosynovial fluid in patients with aseptic loosening of hip prostheses.[64]

Macrophage Activation, Intracellular Signaling, and Transcription Factor NF-kB

A primary response of macrophages to particulate debris is the increased release of TNF-α.[65-68] TNF-α release results in part from exposure of macrophages to particles, which activates the transcription factor NF-κB; this reaction is related to membrane receptor events.[69] Recent studies using a mouse model confirm that activation of NF-κB is of critical importance for the localized resorption of bone in response to particulate debris.[70]

Gene expression increases in inflammatory cells as a result of target gene activation by transcription factors.[71,72] The transcription factors regulate expression of target genes by binding to specific recognition elements in the promotor regions of the gene.[73] DNA binding by transcription factors increases formation of target gene mRNA and protein. The promotor regions of many target genes contain consensus sequences so that a single transcription factor may exert multiple effects.[74,75] Transcription factors exist as inactive cytoplasmic complexes and are considered ubiquitous. In macrophages, transcription factors for many genes are inducible and require de novo protein synthesis through activation of constitutive primary transcription factors by ligand binding, phosphorylation, protease modification, and subunit association/dissociation.[76]

The precise role of NF-κB in the dose- and time-dependent response to particles remains unknown. One major question for the future will be to test the effectiveness of intervention in the NF-κB activation of macrophage cytokine expression as a way of preventing wear debris-activated bone resorption. Anti-inflammatory agents such as glucocorticoids inhibit the action of NF-κB and other transcription factors.[77] Antioxidants such as ascorbate, acetylcysteine, aspirin, and sodium salicylate also inhibit activation of NF-κB.[78-80] The intensity of the cellular response to particulate debris rests in part with a chronic inability of the macrophages to metabolize and/or clear the debris from the site of origin. As a result, macrophages continue to accumulate in the periprosthetic tissue and induce a chronic injury response characterized by secretion of the proinflammatory mediators and bone-resorbing factors. Understanding mechanisms by which NF-κB activates inflammatory gene expression may open new vistas for increasing the longevity of total joint prostheses.

Wear Debris, Macrophage Activation, and Chemokine and Matrix Metalloproteinase Expression

Orthopaedic biomaterial wear particles induce the release of chemokines[81,82] and matrix metalloproteinases[83] from macrophages cultured in vitro. Induction of chemokine release by wear debris generation will attract macrophages to the periprosthetic tissue. Induction of matrix metalloproteinase release by wear debris will be critical to processes involved in the proliferative expansion of the fibrous interfacial membrane at the margins of the prosthesis. These in vitro data are consistent with other observations on matrix metalloproteinase expression in the pathology of a foreign-body reaction.[84] Recent data show that an inducer of matrix metalloproteinase expression, EMMPRIN, is also upregulated in tissue from loosened hip prostheses.[85] The inducer of matrix metalloproteinase expression is associated with tumor cell membrane and involves stimulation of enzyme expression through signaling pathways that influence gene transcription.[86,87] Cathepsin-G, which is able to activate proteases and inactivate metalloproteinase inhibitors as well as influence expression of pro-inflammatory cytokines, has also been identified in macrophages and fibroblasts in periprosthetic tissues.[88] Alteration of the bone surface by these proteases may stimulate osteoclast bone-resorbing activity and may influence recruitment and adhesion of mononuclear phagocyte osteoclast precursors at the bone-implant interface.

The increase in the biologic mediators that occurs in response to particles occurs under a variety of test conditions with direct effects on connective tissue metabolism.[89-91] In some instances, addition of anti-inflammatory cytokines, such as IL-10 and IL-4, can combat the effects of particles on biologic mediator release.[92,93] In addition, the presence of particles may inhibit the function of anti-inflammatory cytokines and further exacerbate osteolysis.[94] However, exacerbation of cytokine release can also occur; with increasing evidence of particle migration,[95] the synergistic interactions of proinflammatory mediators may represent a challenge for in vivo regulatory approaches to blunt the effects of wear debris on bone loss.

Macrophage-Osteoclast Differentiation and Periprosthetic Osteolysis

A second important mechanism relevant to the role of macrophages in implant loosening is revealed in data demonstrating that wear particle-associated macrophages are capable of differentiating into multinucleated cells that exhibit all the phenotypic features of osteoclasts. Osteoclasts are highly specialized multinucleated cells that are uniquely capable of carrying out lacunar resorption. Osteoclasts are formed by fusion of bone marrow-derived mononuclear phagocyte precursors that circulate in the monocyte fraction.[96] Using mouse models of osteoclast formation, it has been shown that monocytes and macrophages, when cocultured with osteoblastic or other bone stromal cells in the presence of 1,25 dihydroxyvitamin D3, [1,25(OH)$_2$D3], can differentiate into mature osteoclasts capable of lacunar bone resorption. Mouse macrophages responding to particulates of all implant biomaterials are capable of differentiating into osteoclasts.[97] There are differences in the extent of osteoclast differentiation (and consequent bone resorption) in association with the macrophage response to different types of biomaterial particles; polyethylene, PMMA, and titanium particles most strongly stimulate macrophage-osteoclast differentiation. Further analysis of PMMA stimulation of macrophage-osteoclast differentiation showed that only PMMA particles containing radio-opaque agents (eg, barium sulphate, zirconium dioxide), resulted in stimulation of macrophage-osteoclast formation and bone resorption.[98]

Using a similar in vitro system, investigators found that large numbers of human osteoclasts can be prepared from human blood monocytes and tissue macrophages. Monocytes and tissue macrophages, cocultured with osteoblast-like cells or other specific bone-derived stromal cells, in the presence of $1,25(OH)_2D3$ and human macrophage CSF, form numerous multinucleated cells that express all the cytochemical and functional phenotypic characteristics of osteoclasts. These cells are positive for tartrate-resistant acid phosphatase, express receptors for calcitonin and vitronectin, and possess the ability to carry out lacunar bone resorption. Because tissue macrophages are derived from monocytes, and monocyte migration is stimulated by the presence of particles, it was not surprising to find that inflammatory foreign body macrophages isolated directly from the pseudomembrane and pseudocapsule around loose arthroplasty components are capable of osteoclast differentiation.[99] In contrast to osteoclast formation by monocytes and macrophages derived from other tissues, the addition of macrophage CSF was found not to be an absolute requirement for osteoclast differentiation by arthroplasty-derived macrophages; this appears to be because sufficient macrophage CSF is produced endogenously in periprosthetic tissues. Correspondingly abundant macrophage CSF has been found in the synovial fluid of prosthetic joints.[100]

Yasuda and associates recently demonstrated that osteoblasts and other bone stromal cells express a novel membrane-bound TNF ligand-like osteoclast differentiation factor termed the receptor activator for nuclear factor κB ligand (RANKL), and that mononuclear phagocyte osteoclast precursors express RANK, a transmembrane signaling receptor for this ligand.[101] RANKL expression by osteoblasts/bone marrow stromal cells and the presence of macrophage CSF are the two main factors required for osteoclast differentiation. A decoy receptor for RANKL—osteoprotegerin (OPG), a member of the TNF receptor superfamily—inhibits osteoclast formation and bone resorption.[102] OPG lacks a transmembrane domain, indicating that it is secreted as a soluble factor that acts to modulate osteoclast formation in an autocrine or paracrine fashion. Human arthroplasty-derived macrophages have been shown to be capable of osteoclast formation in the presence of soluble RANKL alone; this process is inhibited by the addition of OPG.[103] Another group demonstrated that metallic and polymeric particles stimulate expression of RANKL and macrophage CSF as well as other humoral factors that influence osteoclast formation;[104] these particles also stimulate the expression of OPG.

A number of cellular and humoral factors are known to influence RANKL and OPG expression. Osteoclast formation in periprosthetic tissues can effectively be viewed as a balance between the production of these two factors. Various cytokines and growth factors (apart from macrophage CSF) abundant in periprosthetic tissues in aseptic loosening, such as IL-1 and TNF-α, increase OPG mRNA expression by osteoblasts, suggesting that these factors that stimulate osteoclastic bone-resorbing activity appear to act conversely to downregulate osteoclast formation.[105] Prostaglandins such as PGE2 have also been shown to increase RANK production and to decrease OPG release, thus stimulating osteoclast formation and bone resorption. Inflammatory cells, such as T-cells, are present in the arthroplasty membrane and may influence osteoclast differentiation and periprosthetic osteolysis by modulating RANKL expression and OPG production. Macrophages are also capable of producing $1,25(OH)_2D3$, the active metabolite of vitamin D. This has been shown to increase the ratio of RANKL to OPG production and thus osteoclast formation. Recent studies have also highlighted the role of some cytokines (eg, TNF-α, TGF-β, and IL-1) in inducing osteoclast formation both in the presence and absence of RANKL (but not macrophage CSF).

A role for macrophage-osteoclast differentiation in aseptic loosening has been defined by a number of studies that have investigated expression of macrophage and osteoclast markers on mononuclear and multinucleated cells containing wear particles in periprosthetic tissues.[106,107] These studies showed that cells expressing markers of osteoclasts and macrophages are found in periprosthetic tissues, suggesting that osteoclast formation from wear particle macrophages occurs in a stepwise fashion in this tissue location. Crossover of macrophage and osteoclast functions (and evidence of osteoclast membership in the mononuclear phagocyte system) is well illustrated by in vitro studies; investigators have shown that osteoclasts are capable of phagocytosis of wear particles and that this phagocytic function does not abrogate specific osteoclast properties such as lacunar bone resorption and the response to calcitonin (Fig. 1).[108,109] These data support the concept that chronic stimulation of macrophages by particulate debris ultimately involves osteoclastic bone resorption.

Relevance

Therapeutic strategies intended to limit wear particle generation should reduce the extent of the macrophage infiltrate and in this way limit the consequent osteoclast formation and bone-resorbing activity that contribute to periprosthetic osteolysis.

The precise mechanisms by which wear debris leads to osteolysis will ultimately be determined by defining how specific types of particles combine with environmental factors to permit interactions with specific types of cells that then communicate with each other through release of soluble mediators. Work from numerous laboratories using a spectrum of in vitro and in vivo experiments confirm that proinflammatory cytokines are essential to the initiation of events that shift the balance between bone formation

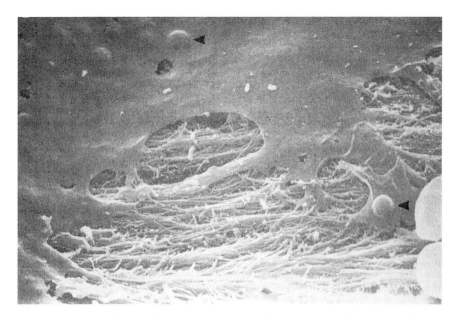

Figure 1 Scanning electron photomicrograph of the edge of an osteoclast that has phagocytosed 1-μm particles (arrows) and is resorbing bone, exposing mineralized collagen fibers. (Reproduced with permission from Wang W, Ferguson DJP, Quinn JMW, Simpson AHRW, Athanasou NA: Osteoclasts are capable of particle phagocytosis and bone resorption. J Pathol 1997;182:92-98.)

and bone resorption. However, the next level of investigation will be to ascertain the precise receptors and signaling pathways that incite changes in gene expression in the target cells.

Although it is generally understood that loss of total joint implant fixation occurs secondary to induction of bone loss by wear debris, no consistent therapeutic approach is available to prevent the process or its consequences in the patient. The significance of the problem derives from the fact that the pathology associated with implant failure affects not only the long-term outcome of the primary arthroplasty but influences outcomes of revision surgery as well. The impact on revision surgery stems from the limitations placed on reconstructive options, increased difficulty of surgery, and the need for bone grafting.

Future Directions for Research

The major consequence of periprosthetic osteolysis is a reduction in available bone to achieve good fixation and implant stability in revision cases. A fundamental understanding of how particulate debris incites the release of biologic mediators, such as the proinflammatory/proresorptive cytokines or cellular recruitment chemokines, will provide important information regarding the development of interventional therapies to prolong the life of a total joint arthroplasty.[110] Successful strategies for improvement of joint replacement surgery may ultimately hinge on interceding in the activation of the final cell type responsible for bone loss, the osteoclast.[111,112] Whether this intervention will occur solely through modulation of biologic mediator activity or through introduction of improved biomaterials that circumvent generation of wear debris remains unknown.[113,114] The development of techniques for the production of human osteoclasts from circulating and tissue mononuclear phagocyte precursors should permit further study of the effect of wear particles on macrophage-osteoclast differentiation. Because osteoclasts have been shown to be capable of particle phagocytosis, the effect of specific wear particles on osteoclastic bone-resorbing activity should also be explored. The effect of specific biomaterial wear particles on stromal cells (eg, osteoblasts) should also be investigated, particularly with regard to the molecular mechanisms that contribute to the promotion of osteoclast formation and bone resorption.

References

1. Amstutz HC, Campbell P, Kossovsky N, Clarke IC: Mechanism and clinical significance of wear debris-induced osteolysis. *Clin Orthop* 1992;276:7-18.
2. Schmalzried TP, Jasty M, Harris WH: Periprosthetic bone loss in total hip arthroplasty: The role of polyethylene wear debris and the concept of the effective joint space. *J Bone Joint Surg Am* 1992;74:849-863.
3. Peters PC Jr, Engh GA, Dwyer KA, Vinh TN: Osteolysis after total knee arthroplasty without cement. *J Bone Joint Surg Am* 1992;74:864-876.
4. Rubash HE, Sinha RK, Shanbhag AS, Kim SY: Pathogenesis of bone loss after total hip arthroplasty. *Orthop Clin North Am* 1998;29:173-186.
5. Kadoya Y, Revell PA, Kobayashi A, al-Saffar N, Scott G, Freeman MA: Wear particulate species and bone loss in failed total joint arthroplasties. *Clin Orthop* 1997;340:118-129.
6. Kim YH, Oh JH, Oh SH: Osteolysis around cementless porous-coated anatomic knee prostheses. *J Bone Joint Surg Br* 1995;77:236-241.
7. Cadambi A, Engh GA, Dwyer KA, Vinh TN: Osteolysis of the distal femur after total knee arthroplasty. *J Arthroplasty* 1994;9:579-594.
8. Wirth MA, Agrawal CM, Mabrey JD, et al: Isolation and characterization of polyethylene wear debris associated with osteolysis following total shoulder arthroplasty. *J Bone Joint Surg Am* 1999;81:29-37.
9. McKellop HA, Campbell P, Park SH, et al: The origin of submicron polyethylene wear debris in total hip arthroplasty. *Clin Orthop* 1995;311:3-20.
10. Devane PA, Bourne RB, Rorabeck CH, MacDonald S, Robinson EJ: Measurement of polyethylene wear in metal-backed acetabular cups: II. Clinical application. *Clin Orthop* 1995;319:317-326.
11. Huk OL, Bansal M, Betts F, et al: Polyethylene and metal debris generated by non-articulating surfaces of modular acetabular components. *J Bone Joint Surg Br* 1994;76:568-574.
12. Huo MH, Betts F, Bogumill GP, Kenmore PI, Hayek RJ, Martinelli TJ: Metallic wear debris in acetabular osteolysis in a mechanically stable cementless total hip replacement: Report of a case. *Orthopedics* 1993;16:1277-1281.
13. Buly RL, Huo MH, Salvati E, Brien W, Bansal M: Titanium wear debris in failed cemented total hip arthroplasty: An analysis of 71 cases. *J Arthroplasty* 1992;7:315-323.
14. Salvati EA, Lieberman JR, Huk OL, Evans BG: Complications of femoral and acetabular modularity. *Clin Orthop* 1995;319:85-93.

15. Bobyn JD, Tanzer M, Krygier JJ, Dujovne AR, Brooks CE: Concerns with modularity in total hip arthroplasty. *Clin Orthop* 1994;298:27-36.

16. Davidson JA, Poggie RA, Mishra AK: Abrasive wear of ceramic, metal, and UHMW-PE bearing surfaces from third-body bone, PMMA bone cement, and titanium debris. *Biomed Mater Eng* 1994;4:213-229.

17. Jasty M, Bragdon C, Jiranek W, Chandler H, Maloney W, Harris WH: Etiology of osteolysis around porous-coated cementless total hip arthroplasties. *Clin Orthop* 1994;308:111-126.

18. Maloney WJ, Peters P, Engh CA, Chandler H: Severe osteolysis of the pelvic in association with acetabular replacement without cement. *J Bone Joint Surg Am* 1993;75:1627-1635.

19. Goodman SB, Huie P, Song Y, et al: Loosening and osteolysis of cemented joint arthroplasties. *Clin Orthop* 1997;337:149-163.

20. Goodman SB, Knoblich G, O'Connor M, Song Y, Huie P, Sibley R: Heterogeneity in cellular and cytokine profiles from multiple samples of tissue surrounding revised hip prostheses. *J Biomed Mater Res* 1996;31:421-428.

21. Jiranek WA, Machado M, Jasty M, et al: Production of cytokines around loosened cemented acetabular components: Analysis with immunohistochemical techniques and in situ hybridization. *J Bone Joint Surg Am* 1993;75:863-879.

22. Santavirta S, Konttinen YT, Hoikka V, Eskola A: Immunopathological response to loose cementless acetabular components. *J Bone Joint Surg Br* 1991;73:38-42.

23. Goldring SR, Schiller AL, Roelke M, Rourke CM, O'Neill DA, Harris WH: The synovial-like membrane of the bone-cement interface in bone lysis. *J Bone Joint Surg Am* 1983;65:575-584.

24. Goldring SR, Jasty M, Roelke MS, Rourke CM, Bringhurst FR, Harris WH: Formation of a synovial-like membrane at the bone-cement interface: Its role in bone resorption and implant loosening after total hip replacement. *Arthritis Rheum* 1986;29:836-842.

25. Kim KJ, Rubash HE, Wilson SC, D'Antonio JA, McClain EJ: A histologic and biochemical comparison of the interface tissue in cementless and cemented hip prostheses. *Clin Orthop* 1993;287:142-152.

26. Kim KJ, Chiba J, Rubash HE: In vivo and in vitro analysis of membranes from hip prostheses inserted without cement. *J Bone Joint Surg Am* 1994;76:172-180.

27. Lee SH, Brennan FR, Jacobs JJ, Urban RM, Ragasa DR, Glant TT: Human monocyte/macrophage response to cobalt-chromium corrosion products and titanium particles in patients with total joint replacements. *J Orthop Res* 1997;15:40-49.

28. Rae T: A study on the effects of particulate metals of orthopaedic interest on murine macrophages in vitro. *J Bone Joint Surg Br* 1975;57:444-450.

29. Maloney WJ, Smith RL, Castro F, Schurman DJ: Fibroblast response to metallic debris in vitro: Enzyme induction, cell proliferation and toxicity. *J Bone Joint Surg Am* 1993;75:835-844.

30. Herman JH, Sowder WG, Anderson D, Appel AM, Hopson CN: Polymethylmethacrylate-induced release of bone-resorbing factors. *J Bone Joint Surg Am* 1989;71:1530-1541.

31. Glant TT, Jacobs J: Response of three murine macrophage populations to particulate debris: Bone resorption in organ cultures. *J Orthop Res* 1994;12:720-731.

32. Glant TT, Jacobs JJ, Molnar G, Shanbhag AS, Valyon M, Galante JO: Bone resorption activity of particulate-stimulated macrophages. *J Bone Miner Res* 1993;8:1071-1079.

33. Horowitz SM, Gautsch TL, Frondoza CG, Riley L: Macrophage exposure to polymethylmethacrylate leads to mediator release and injury. *J Orthop Res* 1991;9:406-413.

34. Lennox DW, Schofield BH, McDonald DF, Riley LH: A histologic comparison of aseptic loosening of cemented, press-fit, and biologic ingrowth prostheses. *Clin Orthop* 1987;255:171-191.

35. Goodman SB, Chin RC, Chiou SS, Schurman DJ, Woolson ST, Masada MP: A clinical-pathologic-biochemical study of the membrane surrounding loosened and non-loosened total hip arthroplasties. *Clin Orthop* 1989;244:182-187.

36. Quinn J, Joyner C, Triffitt JT, Athanasou NA: Polymethylmethacrylate-induced inflammatory macrophages resorb bone. *J Bone Joint Surg Br* 1992;74:652-658.

37. Williams RP, McQueen DA: A histopathologic study of late aseptic loosening of cemented total hip prostheses. *Clin Orthop* 1992;275:174-179.

38. Chiba J, Rubash HE, Kim KJ, Iwaki Y: The characterization of cytokines in the interface tissue obtained from failed cementless total hip arthroplasty with and without femoral osteolysis. *Clin Orthop* 1994;300:304-312.

39. Murray DW, Rushton N: Mediators of bone resorption around implants. *Clin Orthop* 1992;281:295-304.

40. Ohlin A, Johnell O, Lerner UH: The pathogenesis of loosening of total hip arthroplasties: The production of factors by periprosthetic tissues that stimulate in vitro bone resorption. *Clin Orthop* 1990;253:287-296.

41. Sedel L, Simeon J, Meunier A, Villette JM, Launay SM: Prostaglandin E2 level in tissue surrounding aseptic aseptic failed total hips: Effects of materials. *Arch Orthop Trauma Surg* 1992;111:255-258.

42. Ferguson GM, Watanabe S, Georgescu HI, Evans CH: The synovial production of collagenase and chondrocyte activating factors in response to cobalt. *J Orthop Res* 1988;6:525-530.

43. Haynes DR, Rogers SD, Hay S, Pearcy MJ, Howie DW: The differences in toxicity and release of bone-resorbing mediators induced by titanium and cobalt-chromium-aloy wear particles. *J Bone Joint Surg Am* 1993;75:825-834.

44. Murray DW, Rushton N: Macrophages stimulate bone resorption when they phagocytose particles. *J Bone Joint Surg Br* 1991;72:988-992.

45. Felix R, Genolet CL, Lowik C, Cecchini MG, Hofstetter W: Cytokine production by calvariae of osteopetrotic mice. *Bone* 1995;17:5-9.

46. Pollice PF, Silverton SF, Horowitz SM: Polymethylmethacrylate-stimulated macrophages increase rat osteoclast precursor recruitment through their effect on osteoblasts in vitro. *J Orthop Res* 1995;13:325-334.

47. Bataille R, Chappard D, Klein B: The critical role of interleukin-6, interleukin-1 and macrophage colony stimulating factor in the pathogenesis of bone lesions in multiple myeloma. *Int J Clin Lab Res* 1992;21:283-287.

48. Mundy GR: Mechanisms of osteolytic bone destruction. *Bone* 1991;12(suppl 1):S1-S6.

49. Kovacs EJ, DiPietro LA: Fibrogenic cytokines and connective tissue production. *FASEB J* 1994;8:854-861.

50. Oppenheim JJ, Zachariear COC, Mukaida N, Matsushima K: Properties of the novel proinflammatory supergene "intercrine" cytokine family. *Annu Rev Immunol* 1991;9:617-648.

51. Raghow R: The role of extracellular matrix in postinflammatory wound healing and fibrosis. *FASEB J* 1994;8:823-831.

52. Al Saffar N, Revell PA: Interleukin-1 production by activated macrophages surrounding loosened orthopaedic implants: A potential role in osteolysis. *Br J Rheumatol* 1994;33:309-316.

53. al-Saffar N, Mah JT, Kadoya Y, Revell PA: Neovascularisation and the induction of cell adhesion molecules in response to degradation products from orthopaedic implants. *Ann Rheum Dis* 1995;54:201-208.

54. Horowitz SM, Doty SB, Lane JM, Burstein AH: Studies of the mechanism by which the mechanical failure of polymethylmethacrylate leads to bone resorption. *J Bone Joint Surg Am* 1994;75:802-813.

55. Santavirta S, Nordstrom D, Metsarinne K, Konttinen YT: Biocompatibility of polyethylene and host response to loosening of cementless total hip replacement. *Clin Orthop* 1993;297:100-110.

56. Wang JY, Wicklund BH, Gustilo RB, Tsukayama DT: Titanium, chromium and cobalt ions modulate the release of bone-associated cytokines by human monocytes/macrophages in vitro. *Biomaterials* 1996;17:2233-2240.

57. Sabokar A, Rushton N: Role of inflammatory mediators and adhesion molecules in the pathogenesis of aseptic loosening in total hip arthroplasties. *J Arthroplasty* 1995;10:810-816.

58. Shanbhag AS, Jacobs JJ, Black J, Galante JO, Glant TT: Cellular mediators secreted by interfacial membranes obtained at revision total hip arthroplasty. *J Arthroplasty* 1995;10:498-506.

59. Neale SD, Sabokbar A, Howie DW, Murray DW, Athanasou NA: Macrophage colony-stimulating factor and interleukin-6 release by periprosthetic cells stimulates osteo-clast formation and bone resorption. *J Orthop Res* 1999;17:686-694.

60. Xu JW, Konttinen YT, Waris V, Patiala H, Sorsa T, Santavirta S: Macrophage-colony stimulating factor (M-CSF) is increased in the synovial-like membrane of the periprosthetic tissue in the aseptic loosening of total hip replacement (THR). *Clin Rheumatol* 1997;16:243-248.

61. Konttinen YT, Waris V, Xu JW, et al: Transforming growth factor-beta 1 and 2 in the synovial-like interface membrane between implant and bone in loosening of total hip arthroplasty. *J Rheumatol* 1997;24:694-701.

62. Xu JW, Konttinen YT, Li TF, et al: Production of platelet-derived growth factor in aseptic loosening of total hip replacement. *Rheumatol Int* 1998;17:215-221.

63. Al-Saffar N, Khwaja HA, Kodoya Y, Revell PA: Assessment of the role of GM-CSF in the cellular transformation and the development of erosive lesions around orthopaedic implants. *Am J Clin Pathol* 1996;105:628-639.

64. Inomoto M, Miyakawa S, Mishima H, Ochiai N: Elevated interleukin-12 in pseu-dosynovial fluid in patients with aseptic loosening of hip prosthesis. *J Orthop Sci* 2000;5:369-373.

65. Horowitz SM, Purdon MA: Mechanisms of cellular recruitment in aseptic loosening of prosthetic joint implants. *Calcif Tissue Int* 1995;57:301-305.

66. Algan SM, Purdon M, Horowitz SM: Role of tumor necrosis factor alpha in particu-late-induced bone resorption. *J Orthop Res* 1996;14:30-35.

67. Horowitz SM, Luchetti WT, Gonzales JB, Ritchie CK: The effects of cobalt chromi-um upon macrophages. *J Biomed Mater Res* 1998;41:468-473.

68. Xu JW, Konttinen YT, Lassus J, Natah S, Ceponis A, Solovieva S, Aspenberg P, Santavirta S: Tumor necrosis factor-alpha (TNF-alpha) in loosening of total hip replacement (THR). *Clin Exp Rheumatol* 1996;14:643-648.

69. Nakashima Y, Sun DH, Trindade MC, et al: Signaling pathways for tumor necrosis factor-alpha and interleukin-6 expression in human macrophages exposed to titanium-alloy particulate debris in vitro. *J Bone Joint Surg Am* 1999;81:603-615.

70. Schwarz EM, Lu AP, Goater JJ, et al: Tumor necrosis factor-a/nuclear transcription factor-kB signaling in periprosthetic osteolysis. *J Orthop Res* 2000;18:472-480.

71. Papavassiliou AG: Transcription factors. *N Engl J Med* 1995;332:45-47.

72. Zheng ZM, Specter S: Dynamic production of tumour necrosis factor-alpha (TNF-alpha) messenger RNA, intracellular and extracellular TNF-alpha by murine macrophages and possible association with protein tyrosine phosphorylation of STAT1 alpha and ERK2 as an early signal. *Immunology* 1996;87:544-550.

73. Dendorfer U: Molecular biology of cytokines. *Artif Organs* 1996;20:437-444.

74. Grilli M, Jason JS, Lenardo MJ: NF-kB and rel-participants in a multiform transcriptional regulatory system. *Int Rev Cytol* 1993;143:1-62.

75. Nicholson WJ, Slight J, Donaldson K: Inhibition of the transcription factors NF-kappa B and AP-1 underlies loss of cytokine gene expression in rat alveolar macrophages treated with a diffusible product from the spores of Apergillus fumigatus. *Am J Respir Cell Mol Biol* 1996;15:88-96.

76. Baeuerle PA, Henkel T: Function and activation of NF-kB in the immune system. *Annu Rev Immunol* 1994;12:141-179.

77. Scheinman RI, Cogswell PC, Lofquist AK, Baldwin AS: Role of transcriptional activation of IκBa in mediation of immunosuppression by glucocorticoids. *Science* 1995;270:283-286.

78. Wang P, Wu P, Siegel MI, Egan RW, Billah MM: IL-10 inhibits transcription of cytokine genes in human peripheral blood mononuclear cells. *J Immunol* 1994;153:811-816.

79. Wang P, Wu P, Siegel MI, Egan RW, Billah MM: Interleukin (IL)-10 inhibits nuclear factor kB (NFkB) activation in human monocytes. *J Bio Chem* 1996;270:9558-9563.

80. Holschermann H, Durfeld F, Maus U, et al: Cyclosporine a inhibits tissue factor expression in monocytes/macrophages. *Blood* 1996;88:3837-3845.

81. Nakashima Y, Sun DH, Trindade MC, et al: Induction of macrophage C-C chemokine expression by titanium alloy and bone cement particles. *J Bone Joint Surg Br* 1999;81:155-162.

82. Lind M, Trindade MC, Nakashima Y, Schurman DJ, Goodman SB, Smith L: Chemotaxis and activation of particle-challenged human monocytes in response to monocyte migration inhibitory factor and C-C chemokines. *J Biomed Mater Res* 1999;48:246-250.

83. Nakashima Y, Sun DH, Maloney WJ, Goodman SB, Schurman DJ, Smith RL: Induction of matrix metalloproteinase expression in human macrophages by orthopaedic particulate debris in vitro. *J Bone Joint Surg Br* 1998;80:694-700.

84. Imai S, Konttinen YT, Jumppanen M, et al: High levels of expression of collagenase-3 (MMP-13) in pathological conditions associated with a foreign-body reaction. *J Bone Joint Surg Br* 1998;80:701-710.

85. Li TF, Santavirta S, Virtanen I, Kononen M, Takagi M, Konttinen YT: Increased expression of EMMPRIN in the tissue around loosened hip prostheses. *Acta Orthop Scand* 1999;70:446-471.

86. Gou H, Majmudar G, Jensen TC, Biswas C, Toole BP, Gordon MK: Characterization of the gene for human EMMPRIN, a tumor cell surface inducer of matrix metalloproteinases. *Gene* 1998;220:99-108.

87. Lim M, Martinez T, Jablons D, et al: Tumor-derived EMMPRIN (extracellular matrix metalloproteinase inducer) stimulates collagenase transcription through *MAPK* p38. *FEBS Letters* 1998;441:88-92.

88. Takagi M, Konttinen TY, Santavirta S, Kangaspunta P, Suda A, Rokkanen P: Cathespin G and a1-antichymotrypsin in the local host reaction to loosening of total hip prostheses. *J Bone Joint Surg Am* 1995;77:16-25.

89. Maloney WJ, Castro R, Schurman DJ, Smith RL: Effects of metallic debris on adult bovine articular chondrocyte metabolism in vitro. *J Appl Biomater* 1994;5:109-115.

90. Sacomen D, Smith RL, Song Y, Fornasier V, Goodman SB: Effects of polyethylene particles on tissue surrounding knee arthroplasties in rabbits. *J Biomed Mater Res* 1998;43:123-130.

91. Trindade MC, Song Y, Aspenberg P, Smith RL, Goodman SB: Proinflammatory mediator release in response to particle challenge: Studies using the bone harvest chamber. *J Biomed Mater Res* 1999;48:434-439.

92. Pollice PF, Hsu J, Hicks DG, et al: Interleukin-10 inhibits cytokine synthesis in mono-cytes stimulated by titanium particles: Evidence of an anti-inflammatory regulatory pathway. *J Orthop Res* 1998;16:697-704.

93. Trindade MC, Nakashima Y, Lind M, et al: Interleukin-4 inhibits granulocyte-macrophage colony-stimulating factor, interleukin-6 and tumor necrosis factor-alpha expression by human monocytes in response to polymethylmethacrylate particle chal-lenge in vitro. *J Orthop Res* 1999;17:797-802.

94. Sengupta TK, Rakshit DS, Nestor BJ, Sculco TP: Inhibition of interferon-gamma sig-naling by titanium wear debris. *Trans Orthop Res Soc* 2000;25:597.

95. Trindade MC, Lind M, Goodman SB, Maloney WJ, Schurman DJ, Smith RL: Interferon-gamma excerbates polymethylmethacrylate particle-induced interleukin-6 release by human monocyte/macrophages in vitro. *J Biomed Mater Res* 1999;47:1-7.

96. Fujikawa Y, Quinn J, Sabokbar A, McGee JO'D, Athanasou NA: The human mononu-clear osteoclast precursor circulates in the monocyte fraction. *Endocrinology* 1996;139:4058-4060.

97. Pandey R, Quinn J, Joyner C, Triffitt JT, Athanasou NA: Arthroplasty implant biomate-rial particle-associated macrophages differentiate into osteoclastic bone-resorbing cells. *Ann Rheum Dis* 1996;55:388-395.

98. Sabokbar AS, Fujikawa Y, Murray DW, Athanasou NA: Radio-opaque agents in bone cement enhance bone resorption. *J Bone Joint Surg Br* 1997;79:129-134.

99. Sabokbar AS, Fujikawa Y, Neale S, Murray D, Athanasou NA: Human arthroplasty-derived macrophages differentiate into osteoclastic bone-resorbing cells. *Ann Rheum Dis* 1997;56:414-420.

100. Takei I, Takagi M, Ida H, Ogino T, Santavirta S, Konttinen YT: High macrophage-colony stimulating factor levels in synovial fluid of loose artificial hip joints. *J Rheumatol* 2000;27:894-899.

101. Yasuda H, Shima N, Nakagawa N, et al: Osteoclast differentiation factor is a ligand for osteoprotegerin/osteoclastogenesis inhibitory factor and is identical to TRANCE/RANKL. *Proc Natl Acad Sci U S A* 1998;95:3597-3602.

102. Lacey DL, Timms E, Tan H-L, et al: Osteoprotegerin ligand is a cytokine that regu-lates osteoclast differentiation and activation. *Cell* 1998;93:165-176.

103. Itonaga I, Fujikawa Y, Sabokbar A, Athanasou NA: The effect of osteoprotegerin and osteoprotegerin ligand on human arthroplasty macrophage-osteoclast differentiation. *Ann Rheum Dis* 2000;59:26-31.

104. Haynes DR, Rogers SD, Hay S, Pearcy MJ, Howie DW: The differences in toxicity and release of bone-resorbing mediators induced by titanium and cobalt-chromium-alloy wear particles. *J Bone Joint Surg Am* 1993;75:825-833.

105. Hofbauer LC, Khosla S, Dunstan CR, Lacey DL, Boyle WJ, Riggs BL: The roles of osteoprotegerin and osteoprotegerin ligand in the paracrine regulation of bone resorp-tion. *J Bone Miner Res* 2000;15:2-12.

106. Kadoya Y, Al-Saffar N, Kobayashi A, Revell PA: The expression of osteoclast-markers on foreign body giant cells. *Bone Mineral* 1994;27:85-96.

107. Chun L, Yoon J, Song Y, Huie P, Regula D, Goodman S: The characterization of macrophages and osteoclasts in tissues harvested from revised total hip prostheses. *J Biomed Mater Res* 1999;48:899-903.

108. Wang W, Ferguson DJP, Quinn JMW, Simpson AHRW, Athanasou NA: Biomaterial particle phagocytosis by bone-resorbing osteoclasts. *J Bone Joint Surg Br* 1997;79:849-856.

109. Wang W, Ferguson DJP, Quinn JMW, Simpson AHRW, Athanasou NA: Osteoclasts are capable of particle phagocytosis and bone resorption. *J Pathol* 1997;182:92-98.

110. Maloney WJ, Smith RL, Schmalzried T, Huene D, Chiba J, Rubash H: Isolation and characterization of wear debris from the membranes around failed cementless total hip replacements. *J Bone Joint Surg Am* 1995;77:1301-1310.

111. Blaine TA, Rosier RN, Puzas JE, et al: Increased levels of tumor necrosis factor-alpha and interleukin-6 protein and messenger RNA in human peripheral blood monocytes due to titanium particles. *J Bone Joint Surg Am* 1996;78:1181-1192.

112. Sun JS, Lin FH, Hung TY, Chang WH, Liu HC: The influence of hydroxyapatite particles on osteoclast cell activities. *J Biomed Mater Res* 1999;15:311-321.

113. Kim KJ, Hijikata H, Itoh T, Kumegawa M: Joint fluid from patients with total hip arthroplasty stimulates pit formation by mouse osteoclasts on dentin slices. *J Biomed Mater Res* 1998;43:234-240.

114. Jasty M, Jiranek WJ, Harris WH: Acrylic fragmentation in total hip arthroplasty and its biologic consequences. *Clin Orthop* 1992;285:116-128.

Are there host factors that determine/modulate the biologic response to wear particles?

Serial radiographic imaging studies and analysis of periprosthetic tissues retrieved from patients undergoing revision arthroplasty for aseptic loosening of total joint replacements have helped to define host responses to orthopaedic implant wear debris.[1-8] These studies have established that wear particles have the capacity to induce a chronic inflammatory response with histologic features of a foreign body granuloma.[8-17] This tissue response is similar to that seen with other foreign particulate species such as silica, which may be associated with chronic granulomatous diseases.[12] The patterns of response to different particulate species are not uniform. Responses to wear particles vary with differing composition and physical chemical properties, and there is a hierarchy in the capacity of different particulate species to induce an inflammatory response.[13-18] What has been more difficult to establish, however, is whether individual patients exhibit differential capacities to respond to implant wear and what factors might underlie this potential diversity in host responses.

When considering the mechanisms responsible for the variability in individual host responses to wear debris, the contribution of individual factors that could potentially modulate the pattern of the host response must be carefully identified. Cellular and tissue responses to wear particles are not limited to the capacity of an individual to mount a biologic response to particles. The pattern and magnitude of the host response may also be affected by biomechanical and physical factors, such as the rates of wear particle generation and the influences of the pattern of gait and load transfer on the physical properties of the wear particles and their rate and tissue distribution at the implant-bone interface. Schmalzried and associates[19] have observed that zones of osteolysis associated with particle-induced granulomatous inflammation tend to occur in regions that are in contact with synovial fluid containing the particulate wear debris. Synovial fluid can penetrate periprosthetic sites that are distant from the actual joint cavity, such as regions associated with fragmentation of the cement mantle, and redistribute particles to other locations.[20] Localized accumulation of particles at these distant sites may produce a local tissue response leading to the development of focal osteolysis.[3,21-24] In this case, the pattern of host response is not necessarily dependent on the individual's unique capacity to mount an inflammatory reaction to the particles but rather to the topographical distribution of potential pathways by which particles can penetrate regions away from the joint cavity. These pathways are influenced by the design of the implant, the surgical techniques associated with the original implant fixation, and the effects of gait on intra-articular pressures and fluid flow within the joint cavity and adjacent periprosthetic tissues.[25,26] Thus, a relatively limited quantity of wear particles could move to a restricted region and accumulate locally in high concentrations, producing a disproportionately large area of inflammation and osteolysis. Under these circumstances, it could be incorrectly concluded that the patient had enhanced biologic reaction to the wear particles, when in fact the major factor responsible for the extensive focal osteolysis was the selective concentration of the particles at a single site.

There is generally a positive correlation of the incidence and magnitude of implant wear with high rates of osteolysis. In this paradigm, the presence of osteolysis has been used as a surrogate for quantifying the adverse tissue response to wear particles. However, osteolysis represents only the final consequence of a complex sequence of biologic events that eventually leads to induction of a localized disturbance in bone remodeling characterized by enhanced bone resorption. The initial stage of host-particle interaction includes particle recognition by host defense systems, recruitment of inflammatory cells to the tissue site, and cell-particle interactions leading to the formation of the tissue granuloma. In turn, the presence of the granulomatous inflammatory response at skeletal sites leads to increased bone resorption (Table 1). It is likely that each stage in the evolution of the inflammatory response and the subsequent bone remodeling defect are modulated in part by host factors that are determined by genetically controlled processes.

To some extent, the host's response to wear particles is influenced by the nature and magnitude of the inflammatory response. Analysis of tissues retrieved from failed implants associated with wear debris have established that this tissue reaction has the characteristic histopathologic features of a granuloma. Granulomas have been defined as a distinct host response mechanism that is activated in response to materials or substances that have resisted destruction by components of the acute inflammatory response. Granulomas provide the host with a defense system in which macrophages, the principal cell type within the granuloma, recruit fibroblast-like cells that deposit a collagenous matrix and sequester the foreign material that cannot be removed or degraded. Granulomas have been classified as immune or nonimmune. Immune granulomas characteristically are associated with antigenic materials derived from bacteria or fungi and exhibit features of

Table 1   Biologic events associated with host responses to particles

Particle recognition
Recruitment of inflammatory cells
Particle-cell interaction
Particle-induced release of proinflammatory products (eg, cytokines, proteases, etc)
Granuloma formation
Skeletal tissue response to granulomatous inflammation

delayed hypersensitivity manifest by the infiltration of T and B lymphocytes. Although foreign inorganic particulate materials are not usually associated with the induction of immune granuloma, one exception is manifest in individuals with silicosis in whom intrapulmonary granuloma may be infiltrated with T and B cells. These individuals may exhibit hypergammaglobulinemia and develop autoantibodies consistent with T and B cell activation. Silicosis thus serves as a model in which certain susceptible individuals, presumably determined by genetic factors, exhibit a unique hypersensitivity response to a foreign inorganic particulate material or to antigenic materials that are adsorbed onto the surface of the particles.

Analysis of retrieved periprosthetic tissues from patients with wear debris-induced osteolysis and aseptic loosening has revealed evidence of T cell infiltration in some cases, but it is not clear that the presence of these cells is specific evidence of hypersensitivity. Although these findings would suggest that a majority of patients do not develop particle-induced hypersensitivity, there is evidence that some patients may exhibit features of hypersensitivity directed either to some component of the implant material itself or to adsorbed products that possess antigenic activity.[27] Lalor and associates[28] reported significant lymphocyte infiltration around titanium-based implants in selected patients. Based on the presence of positive metal skin testing, they speculated that hypersensitivity to the metal had contributed to the adverse tissue reaction. Additional studies have demonstrated the development of metal sensitivity to metal wear debris or corrosion products also exists, but the actual numbers of patients in whom this plays a significant role in the pathogenesis of the implant failure appears to be small.[27,29-31]

Individuals with pre-existing rheumatoid arthritis (RA) demonstrate significant evidence of immune granulomatous responses to implant wear particles.[32] Analysis of retrieved tissues from these patients demonstrated heavy infiltration of periprosthetic tissues with lymphocytes. Caplan and associates[33] described a similar immune granulomatous reaction in the lungs of patients with RA who were exposed to particulate coal dust. Although the specific factors responsible for this unique cellular response are not clear, these findings provide direct evidence that there are distinct subsets of individuals who exhibit differential patterns of response to foreign particulate materials. The response patterns appear to be related to underlying host factors that are determined by individual genetic predispositions.

In attempting to dissect the specific factors underlying the potential differential host responses to wear particles, experimental approaches must permit examination of the specific biological events associated with each of the individual stages of host-particle interaction. Some individuals might process or internalize subcellular particulate species more efficiently, while others may have enhanced capacity to respond to the cell-particle interaction or phagocytic activation with enhanced release of cytokines or other products implicated in the development of granulomatous inflammation. With respect to the skeletal response to granulomas, individuals probably vary in their inflammatory reaction and subsequent bone remodeling. Each step in the pathogenesis of the host response is regulated by cellular and biochemical processes that are determined by specific genetic factors. Insight into these factors could facilitate the identification of individuals who would be more likely to respond adversely to wear particles. This in turn could lead to the development of more effective treatment regimens that specifically target the biologic processes responsible for the enhanced response to wear particles.

Genetic analysis has been used to study a variety of inflammatory disorders, and provides convincing evidence that genetic factors play a critical role in determining the pattern and outcome of the host response to proinflammatory stimuli. Analysis of patients with RA represents an excellent model for exploring the role genetic factors play in determining not only disease predisposition but also severity of systemic inflammatory manifestations as well as bone and cartilage destruction.[34-37] Data derived from twin studies indicate that if one member of a monozygotic twin pair develops RA, the risk that the sibling will develop RA is greater than 30%. Analysis of the *HLA-DR* locus (a chromosomal region containing genes that regulate an individual's immune response) in patients with RA demonstrated that 80% of Caucasian subjects with RA express specific *DR1* or *DR4* genes that share a region of highly similar nucleotide sequence designated the "shared epitope". The presence of these specific DRβ genetic subtypes (alleles) may confer a unique propensity to the development of more aggressive disease characterized by bone erosion and cartilage and joint destruction.

Genetic linkage studies indicate that the *HLA* locus accounts for only one component of the RA disease association; other unidentified genetic loci are also involved.[34,38] Identification of the genetic factors predisposing to RA has been a challenging undertaking because RA is a clinically heterogeneous disorder that is almost certainly related to the interplay of complex genetic influences. Thus, similar to disorders such as diabetes, osteoporosis, and hypertension, RA is not likely to be a monogenic disorder. Adding to the challenge, the genetic complexity in a disorder such as RA may also lead to a situation in which different genetic combinations may give rise to similar patterns of disease expression.

The recent development of high through-put screening approaches for large-scale, genome-wide analysis has enabled researchers to begin dissecting the underlying genetic factors responsible for complex genetic disorders such as RA. Technologies that permit identification of single nucleotide polymorphisms (SNPs), which are single base differences in the DNA sequence observed among individuals in a population, have been particularly useful.[39] These variations in DNA sequence are common (average frequency 1 per 1,000 base pairs) and are present throughout the human genome.

SNP-based screening and association studies can be used for genome scanning, usually within sibling pairs, in order to identify association between a disease phenotype and sharing of chromosomal DNA sequence identified by the presence of conserved SNPs. Investigators look for linkage disequilibrium, which occurs when haplotype combinations of alleles (detected by SNPs) occur more frequently in a disorder than would be expected from random association.[39] This approach has been used to identify the association of polymorphisms in the *ApoE* gene in a subset of patients with Alzheimer's disease. A second approach for SNP-based screening and association studies is called the candidate gene approach. With this approach, attention is focused on specific genes that have been implicated in clinical or laboratory research on the pathogenesis of a disorder. One application would be the investigation of the association of polymorphisms in the tumor necrosis factor-α (TNF-α) or interleukin-1 (IL-1) genes, cytokines implicated in the proinflammatory and destructive events in inflammatory disorders, with RA.[35,37] Additional studies have focused on the corticotrophin-releasing hormone gene, because of the pivotal role this gene plays in regulating systemic glucocorticoid responses to stress.[34] Although such studies implicating these genes in disease pathogenesis or association await validation, initial results suggest that they hold promise for providing useful insights into the genetic basis of complex human diseases. Because many of the polymorphisms are localized to the regions of genes associated with gene regulation, the presence of alterations in DNA sequence could modulate the levels of gene expression leading to dysregulated host responses. In this respect, analysis of polymorphisms may also yield insights in the mechanisms underlying gene regulation.

These approaches have not been applied to the analysis of host responses to wear particles. Despite their potential for providing insights into these biologic processes, the individual biologic events associated with host responses to wear particles must be carefully dissected before attempting these analyses. Certain biologic processes may lend themselves especially well to this type of analysis. For example, some individuals may demonstrate enhanced capacity to release proinflammatory cytokines in response to phagocytic activation. In this case, selection of candidate genes for study might include IL-1 or TNF-α, which have been shown to be involved in the pathogenesis of periprosthetic osteolysis in analyses of retrieved tissues and in vivo and in vitro cell culture models. However, because of the multiplicity of factors that are ultimately responsible for the development of osteolysis, it may be very difficult to firmly establish the role of individual genetic factors in host responses to wear particles. With additional research, it may be possible to develop a gene profile that identifies individuals at high risk for adverse responses.

Experimental strategies and appropriate technologies are now available that will permit high through-put screening for the identification of target genes that are affected by host responses to wear particles. These include approaches to characterize the expression profiles of genes (transcriptional profiling) or their encoded proteins (proteomics). Transcriptional profiling, employing microarray technology, was recently used to examine the effects of wear particles on the regulation of large numbers of genes and gene families.[40] Results indicate that macrophage activation by wear particles induces distinct patterns of gene response. Because of the large numbers of genes affected by the cell treatment, data analysis is formidable. Characterization of responses requires integration with appropriate bioinformatics resources that permit discrimination of relevant changes from background noise and reduction in the influence of artifacts related to the technology. Another issue is the limitation to analysis of RNA levels that may correlate poorly with the actual protein levels of the encoded genes.

Relevance

Clinical assessment of patients exposed to orthopaedic implant wear suggests that host responses to wear particles are heterogeneous. Identification of the specific factors responsible for these differential responses has significant clinical importance because of the role of wear particles in the pathogenesis of aseptic implant loosening. Technologies are now available for identifying the genetic factors that determine the pattern of host responses in inflammatory disorders. Successful application of these technologies to define the role of individual genetic factors that regulate host responses to implant wear will depend on the availability of carefully selected patient groups in whom the independent effects of other factors affecting host responses, such as the physical-chemical properties of the particles and the influences of biomechanical factors, have been appropriately controlled. Identification of the specific genetic factors that determine the pattern of host responses will not only provide insights into pathogenic mechanisms regulating responses to wear particles but also could lead to improved approaches for preventing periprosthetic osteolysis and implant failure.

Future Directions for Research

Research efforts should focus on identifying and describing unique patients or cohorts with an immunologic basis for osteolysis; developing effective treatments that specifically target biologic processes responsible for osteolysis; and the use of genomic and proteonic analysis to identify patients at risk for developing osteolysis.

References

1. Charnley J: The reaction of bone to self-curing acrylic cement. *J Bone Joint Surg Br* 1970;52:340-353.
2. Fornasier VL, Cameron HU: The femoral stem/cement interface in total hip replacement. *Clin Orthop* 1976;116:248-252.
3. Harris WH, Schiller AL, Scholler JM, Freiberg RA, Scott R: Extensive localized bone resorption in the femur following total hip replacement. *J Bone Joint Surg Am* 1976;58:612-618.
4. Mirra JM, Amstutz HC, Matos M, Gold R: The pathology of the joint tissues and its clinical relevance in prosthesis failure. Clin Orthop 1976;117:221-240.
5. Mirra JM, Marder RA, Amstutz HC: The pathology of failed total joint arthroplasty. *Clin Orthop* 1982;170:175-183.

6. Vernon-Roberts B, Freeman MAR: Morphological and analytical studies of the tissues adjacent to joint prostheses: Investigations into the causes of loosening prostheses, in Schaldach M, Hohmann D (eds): *Advances in Artificial Hip and Knee Joint Technology.* Berlin, Springer-Verlag, 1976, p 148-186.

7. Willert HG, Ludwig J, Semlitsch M: Reaction of bone to methacrylate after hip arthroplasty. *J Bone Joint Surg Am* 1974;56:1368-1382.

8. Willert HG, Semlitsch M: Tissue reactions to plastic and metallic wear products of joint endoprostheses, in Gschwend N, Debrunner HU (eds): *Total Hip Prosthesis.* Baltimore, MD, Williams and Wilkins, 1976, pp 205-239.

9. Goldring SR, Schiller AL, Roelke M, Rourke CM, O'Neill DA, Harris WH: The synovial-like membrane at the bone-cement interface in loose total hip replacements and its proposed role in bone lysis. *J Bone Joint Surg Am* 1983;65:575-584.

10. Shanbhag AS, Jacobs JJ, Black J, Galante JO, Glant TT: Cellular mediators secreted by interfacial membranes obtained at revision total hip arthroplasties. *J Arthroplasty* 1995;10:498-506.

11. Willert HG, Semlitsch M: Reactions of the articular capsule to wear products of artificial joint prostheses. *J Biomed Mater Res* 1977;11:157-164.

12. Summerton J, Hoenig S: The mechanism of hemolysis by silica and its bearing on silicosis. *Exp Mol Pathol* 1977;26:113-128.

13. Cohen J: Assay of foreign-body reaction. *J Bone Joint Surg Am* 1959;41:152-166.

14. Rae T: Action of wear particles from total joint replacement prostheses on tissues, in Williams DF (ed): *Biocompatibility of Implant Materials.* London, UK, Sector Publishing, 1976, pp 55-599.

15. Rae T: The haemolytic action of particulate metals. *J Pathol* 1978;125:81-89.

16. Rae T: The biological response to titanium and titanium-aluminum-vanadium alloy particles: I. Tissue culture studies. *Biomaterials* 1986;7:30-36.

17. Shanbhag AS, Hasselman CT, Jacobs JJ, Rubash HE: Biologic response to wear debris, in Callaghan JJ, Rosenberg AG, Rubash HE (eds): *The Adult Hip.* Philadelphia, PA, Lippincott-Raven Publishers, 1998, pp 279-288.

18. Raut VV, Siney PD, Wroblewski BM: Cemented Charnley revision arthroplasty for severe femoral osteolysis. *J Bone Joint Surg Br* 1995;77:362-365.

19. Schmalzried TP, Kwong LM, Jasty MJ, et al: The mechanism of loosening of cemented acetabular components in total hip arthroplasty. *Clin Orthop* 1992;274:60-78.

20. Anthony PP, Gie GA, Howie CR, Ling RSM: Localised endosteal bone lysis in relation to the femoral components of cemented total hip arthroplasties. *J Bone Joint Surg Br* 1990;72:971-979.

21. Aspenberg P, Herbertsson P: Periprosthetic bone resorption. *J Bone Joint Surg Br* 1996;78:641-646.

22. Carlsson AS, Gentz CF, Linder L: Localized bone resorption in the femur in mechanical failure of cemented total hip arthroplasties. *Acta Orthop Scand* 1983;54:396-402.

23. Gross TP, Lennox DW: Osteolytic cyst-like area associated with polyethylene and metallic debris after total knee replacement with an uncemented vitallium prosthesis. *J Bone Joint Surg Am* 1992;74:1096-1101.

24. Jasty MJ, Floyd WE, Schiller AL, Goldring SR, Harris WH: Localized osteolysis in stable, non-septic total hip replacement. *J Bone Joint Surg Am* 1986;68:912-919.

25. Schmalzried TP, Guttman D, Grecula M, Amstutz HC: The relationship between the design, position and articular wear of acetabular components inserted without cement and the development of pelvic osteolysis. *J Bone Joint Surg Am* 1994;76:677-688.

26. Zicat B, Engh CA, Gokcen E: Patterns of osteolysis around total hip components inserted with and without cement. *J Bone Joint Surg Am* 1995;77:432-439.

27. Evans EM, Freeman MAR, Miller AJ, Vernon-Roberts B: Metal sensitivity as a cause of bone necrosis and loosening of the prosthesis in total joint replacement. *J Bone Joint Surg Br* 1974;56:626-642.

28. Lalor PA, Revell PA, Gray AB, Wright S, Railton GT, Freeman MAR: Sensitivity to titanium (A cause of implant failure?). *J Bone Joint Surg Br* 1991;73:25-28.

29. Merritt K, Rodrigo JJ: Immune response to synthetic materials sensitization of patients receiving orthopaedic implants. *Clin Orthop* 1996;326:71-79.

30. Santavirta S, Konttinen YT, Bergroth V, Gronblad M: Lack of immune response to methyl methacrylate in lymphocyte cultures. *Acta Orthop Scand* 1991;62:29-32.

31. Wooley PH, Fitzgerald RH Jr, Song Z, et al: Proteins bound to polyethylene components in patients who have aseptic loosening after total joint arthroplasty. *J Bone Joint Surg Am* 1999;81:616-623.

32. Goldring SR, Wojno WC, Schiller AL, Scott RD: In patients with rheumatoid arthritis the tissue reaction associated with loosened total knee replacements exhibits features of a rheumatoid synovium. *J Orthop Rheum* 1988;1:9-21.

33. Caplan A, Payne RB, Whithey JL: A broader concept of Caplan's syndrome related to rheumatoid factors. *Thorax* 1962;17:205-212.

34. Fife MS, Fisher SA, John S, et al: Multipoint linkage analysis of a candidate gene locus in rheumatoid arthritis demonstrates significant evidence of linkage and association with the corticotropin-releasing hormone genomic region. *Arthritis Rheum* 2000;43:1673-1678.

35. Mattey DL, Hajeer AH, Dababneh A, et al: Association of giant cell arthritis and polymyalgia rheumatica with different tumor necrosis factor microsatellite polymorphisms. *Arthritis Rheum* 2000;43:1749-1755.

36. Nieto A, Caliz R, Pascual M, Mataran L, Garcia S, Martin J: Involvement of $Fc\gamma$ receptor IIIA genotypes in susceptibility to rheumatoid arthritis. *Arthritis Rheum* 2000;43:735-739.

37. Shibue T, Tsuchiya N, Komata T, et al: Tumor necrosis factor α 5'-flanking region, tumor necrosis factor receptor II, and HLA-DRB1 polymorphisms in Japanese patients with rheumatoid arthritis. *Arthritis Rheum* 2000;43:753-757.

38. Boehnke M: A look at linkage disequilibrium. *Nat Genet* 2000;25:246-247.

39. Gray IC, Campbell DA, Spurr NK: Single nucleotide polymorphisms as tools in human genetics. *Hum Mol Genet* 2000;9:2403-2408.

40. Shanbhag AS, Cho DR, Choy BK, et al: The transcriptional response program of human monocyte activation by polyethylene. *Trans Orthop Res Soc* 2000;25:52.

What specific features of wear particles are most important in determining the adverse biologic reactions?

Wear particle size, concentration, material, and form are the features that appear to be significant in the adverse biologic reactions associated with failed total joint replacements. The key cell type in these adverse reactions is the tissue macrophage or histiocyte (Fig. 1); these cells are the major source of the inflammatory and potentially osteolytic cytokines. However, fibroblasts and osteoblasts are also exposed to wear particles and can produce cytokines that affect osteoclasts.

Size of Particles

Particle phagocytosis plays a key role in macrophage activation and hence the size of the particle is significant. Histologic observations of tissues around failed total joint replacements frequently note an abundance of enlarged macrophages, often with visible particles within their cytoplasm (Fig. 2). The true amount of wear particles in these cells was not fully appreciated until techniques to visualize and characterize intracellular particles were applied to joint tissues (Figs. 3 and 4). It became apparent that the majority of wear particles in these tissues were submicron to nanometer in size (Figs. 5 and 6), below the level of resolution of most light microscopes.[1-4]

The production of inflammatory cytokines by the cells within tissues adjacent to failed total hip prostheses was verified by a number of research groups (Fig. 7).[5-10] The association between ingested wear particles and cytokine production led to extensive research into the particle parameters and cellular mechanisms that are involved.

The critical size range for phagocytosis-induced macrophage activation by wear particles has been estimated to be from 0.2 to 10 μm.[11-14] A subcutaneous rat air-pouch model was used to assess the in vivo effects of the size of polymethylmethacrylate (PMMA) particles on the acute inflammatory response. For a given mass or dose, the small (< 20 μm) particles elicited a significantly greater inflammatory reaction than the large (50 to 350 μm) particles, as expressed by the release of tumor necrosis factor (TNF), neutral metalloprotease, and prostaglandin E2 (PGE2).[15] Particles of PMMA small enough to be phagocytized (< 12 μm) resulted in the increased production of TNF by macrophages in a cell culture model.[16] To further define the

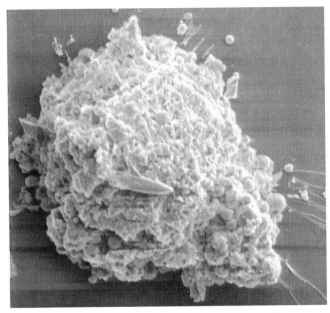

Figure 1 Scanning electron micrograph of a macrophage isolated from tissue around a failed total hip replacement. (2500 ×)

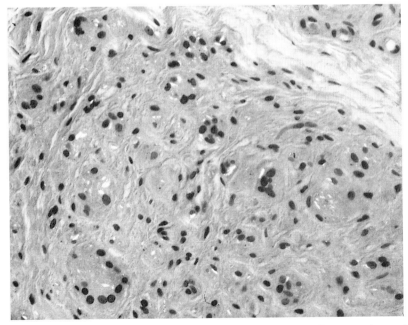

Figure 2 Light micrograph of tissue from a failed total hip replacement shows abundant macrophages containing polyethylene wear particles that are seen as white fibers and flakes under polarized light. (Haematoxylin and eosin)

95

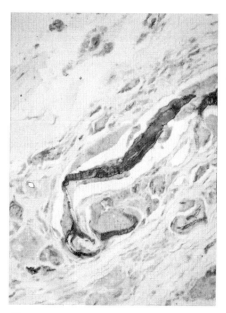

Figure 3 *Light micrograph of tissue from a failed polyethylene socket shows several large strands of polyethylene stained by oil red O. (Toluidine blue, oil red O stain)*

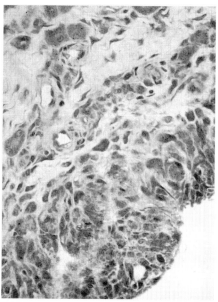

Figure 4 *Typical histology of the interfacial membrane tissue surrounding a failed total hip replacement. Macrophages with oil red O-stained cytoplasm indicate intracellular polyethylene particles too small to be seen readily by light microscopy. (Toluidine blue, oil red O stain)*

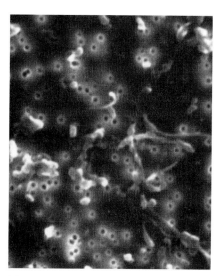

Figure 5 *Scanning electron micrograph of polyethylene particles isolated from the tissue in Figure 4. The background holes are 0.2 microns in diameter. (20,000 ×)*

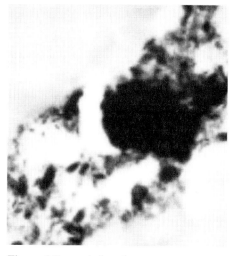

Figure 6 *Transmission electron micrograph of a group of cobalt chromium and chromium oxide particles isolated from tissue around a failed metal-metal total hip replacement. (31,000 ×)*

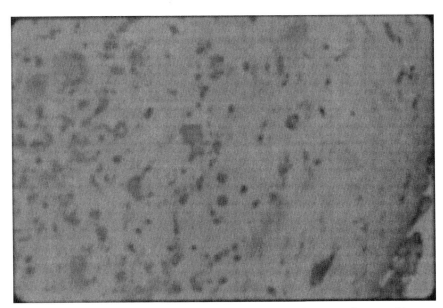

Figure 7 *Light micrograph of tissue from a failed total hip replacement shows positive staining for IL-6.*

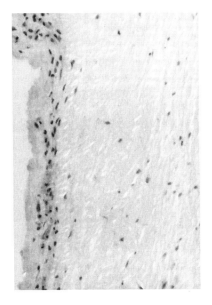

Figure 8 *Light micrograph of capsule tissue from a metal-metal total hip replacement after 19 months shows a small number of macrophages arranged along the implant facing edge. (Polarized, haematoxylin and eosin)*

effects of particles size, the in vitro response of murine peritoneal macrophages to polyethylene particles of definitive size distributions (mean sizes of 0.21, 0.49, 4.3, 7.2, and 88 µm) was measured by IL-6, IL-β and TNF-α production. Particles of 0.49, 4.3, and 7.2 µm produced the highest levels of these cytokines.[11]

Another important aspect of particle size is the total surface area of the ingested particles. Shanbhag and associates[13] introduced the concept of surface area ratio to study the cellular responses of cells to wear particles in a standardized manner. Using a monocyte/macrophage cell line, increasing IL-1 levels were detected at higher surface area ratios with titanium particles. At a constant surface area ratio, the smaller particles (0.15 and 0.45 µm) were consistently less inflammatory than the larger particles (1.76 µm). This may account for the observations made by Doorn and associates of tissues around metal-metal total hip replacements (Fig. 8).[17] Although the number of particles was higher because of their predominantly nanometer size, the tissues were consistently less inflamed than tissues around metal-polyethylene hips, which had fewer but larger particles.

Particle contact with the macrophage surface membrane initiates intracellular signaling events that lead to cytokine production; phagocytosis is not necessary for the induction of TNF-α and IL-1 by macrophages exposed to particles.[18,19] Particle size is still important to this process, however, because membrane recognition may not occur above or below a certain size range.

The influence of hydroxyapatite particles of differing sizes (0.5 to 841 μm) on bone cell activities was studied in vitro by Sun and colleagues.[20] The inhibitory effects of hydroxyapatite particles on osteoblast cell cultures were mediated by the increased synthesis of PGE2 concentration, which was more significant and persisted longer in the smallest particle group (0.5 to 3.0 μm).

Concentration of Particles

Several clinical studies have shown a strong association between polyethylene wear rate and the occurrence of osteolysis or loosening. In a group of 235 Charnley low-friction arthroplasties, the prevalence of osteolysis (33 hips, 14%) and of acetabular and femoral component loosening and revision rose significantly with increasing wear. The wear rate of revised components was twice that of surviving components. For every additional millimeter of wear, the risk of acetabular revision in any 1 year increased by 45%; for the femur, risk increased by 32%.[21] Similarly, femoral-head penetration patterns were examined at a minimum of 10 years in 48 primary total hip arthroplasties using an Arthropor acetabular cup with a 32-mm cobalt chrome (CoCr) alloy ball.[22] Osteolysis at 10 years was strongly associated with increasing wear rates. Osteolysis did not develop in any of the nine hips with a wear rate of less than 0.1 mm per year, but developed in all eight hips with a rate exceeding 0.3 mm per year. These observations suggest that the amount of wear debris is influential in determining the tissue response.

Kobayashi and associates[23] studied the size, shape, and concentration of polyethylene wear particles obtained from peri-implant tissues of patients having revision surgery for aseptically loosened total joint replacements. The concentration of polyethylene particles in six areas with osteolysis was significantly higher than that in 12 areas without osteolysis. There were no significant differences between the size and shape of the particles in these two groups. The authors concluded that the most critical factor in the pathogenesis of osteolysis is the concentration of polyethylene particles accumulated in the tissue. These findings were supported by an in vitro study using macrophage-like cells cultured in the presence of polyethylene and titanium particles on a rotation device that simulates the contact between cells and particles that occurs in vivo and encourages particle-cell contact and phagocytosis.[24] Low concentrations of particles caused little inflammatory mediator release, but higher levels of particles resulted in maximal release of mediators.

Because it is difficult to isolate the effects of particle size and concentration, numerous studies have examined these in combination. Primary human monocyte/macrophages were cultured with different sizes, concentrations, volumes, and surface areas of PMMA particles.[25] The release of cytokines, PGE2, and hexosaminidase depended on the size, concentration, surface area, and volume of the phagocytosable particles. Catelas and associates[26] used flow cytometric analysis to study the macrophage response to different sizes and concentrations of ceramic and polyethylene particles. Phagocytosis increased with concentration for small particles (< 2 μm); with larger particles (up to 4.5 μm), phagocytosis seemed to reach a plateau independent of size and concentration. This suggested a saturation of phagocytosis possibly dependent on the overall particle volume ingested. TNF-α release was found to increase with particle concentration. Other studies showed that apoptosis of a macrophage cell line was induced by particles of ceramic and polyethylene in a size- and concentration-dependent manner.[27]

When the concentration of metal particles increases, some of the cell or tissue responses may be related to toxic effects. Rae[28] incubated human synovial fibroblasts with various preparations of metals for periods up to 18 days. Cultures exposed to particulate pure metals were poisoned by cobalt and vanadium but were unaffected under the same conditions by nickel, chromium, molybdenum, titanium, or aluminum. Haynes and colleagues[29] compared the in vitro response of human monocytes to similar concentrations of 1-μm particles of cast and forged CoCr, stainless steel, and titanium alloy (TiAlV). There was no difference in biologic effects of the two cobalt chromes and stainless steel alloys; all were toxic to cells, while titanium alloy was not.

Maloney and associates[30] found chondrocyte viability to be influenced by the type and concentration of metal particles. Cobalt was toxic to chondrocytes at all particle concentrations (0.83% to 0.000083%, v/v); however, the chromium, titanium, and titanium-aluminum alloy particles only affected chondrocyte viability at high concentrations. These findings are consistent with studies of two human osteoblast-like cell lines exposed to particulate cobalt, chromium, and cobalt-chromium alloy at concentrations of 0, 0.01, 0.1, and 1.0 mg/ml[31] as well as studies of primary lines of fibroblastic cells from newborn rats exposed to powders of cobalt-chrome-molybdenum alloy and its main constituents.[32]

Material

The assessment of the role of material in the cellular response to wear particles has often been confounded by comparisons of particulate materials of dissimilar size and shape. While this problem has not been completely overcome, better availability of manufactured particles with controlled size and shape characteristics have minimized the differences in size, shape, and surface area between materials and have highlighted a clear effect of particle material in in vitro studies. This is particularly clear when the material in question has potential toxicity, such as cobalt chrome.[33-36] When used in non-toxic doses, differences in cytokine production by macrophages have been reported between titanium and cobalt chrome,[37] between titanium particles from differing alloys,[35] and between titanium and polyethylene.[38]

Differences have also been demonstrated with other cell types. Confluent primary human osteoblasts and MG63 osteoblast-like cells were incubated in the presence of particles of commercially pure titanium, titanium-aluminum-vanadium (Ti-6Al-4V) alloy, cobalt chrome alloy, and polyethylene of equivalent size.[39] All three types of particles were phagocytosed and had different effects on the cells. The effect on cell number was dependent on the chemical composition of the particles; titanium alloy and CoCr caused a dose-dependent increase, while commercially pure (cp) Ti particles had a biphasic effect with a maximal increase in cell number observed at the 1:10 dilution. PGE2 production was increased by all particles, but the magnitude of the effect was particle-dependent: CoCr > cpTi > Ti alloy.

Human bone-derived cells were incubated with equivalent concentrations of particles of Ti-6Al-4V, cpTi, and CoCr alloy of 0.5 to 3.0 μm. Mediators released from cells cultured with Ti-6Al-4V particles induced human bone-derived cells to express higher levels of collagen type I, osteopontin, bone sialoprotein, and osteocalcin than those cultured with cpTi, CoCr, and unstimulated controls.[40]

The effect of polyethylene (PE) particle chemistry on human monocyte-macrophage function was studied in monocyte-derived macrophages seeded onto glass coverslips coated with a type 1 collagen matrix embedded with virgin high density (HD) PE and chemically modified HDPE. The modified HDPE had been impregnated with 435-ppm levels of $CoCl_2$ and oxidized for up to 120 hours by heat. The cells were cultured for 96 hours. A significant elevation of IL-1 secretion was observed after initial exposure to virgin HDPE particles compared with controls ($P = 0.001$). IL-1 secretion was also elevated in the low oxidized particle groups ($P = 0.001$), while the highly oxidized particles were no different than controls. Secretion of both IL-6 and TNF-α were significantly elevated by the low oxidized HDPE particles; the virgin and highly oxidized groups showed no difference.[41] The introduction of new polyethylenes may result in particles with different surface chemistry than the previously characterized particles. These changes should be characterized to determine if they alter the particle protein adsorption profiles and subsequent biologic reactivity.

In a comparison of the inflammatory response to particulate PMMA debris with and without barium sulfate in the rat subcutaneous pouch model, the inflammatory response to PMMA containing $BaSO_4$ was greater than the response to plain PMMA particles of similar size.[42] Similar results were reported with a macrophage cell line model.[43]

Another important aspect of particle material is its degradability. Because PE particles are highly resistant to chemical attack, they persist in the tissues. Metal and, to a lesser extent, ceramic particles will degrade and dissolve over time. This degradation process can alter the cellular interaction with the material from particle-based to ion-based.

Form

Although size, concentration, and material appear to be the most significant aspects of wear particles affecting the biologic responses in the joint, several other variables should be discussed; the term "form" is used here to differentiate these from the parameters discussed above. One such variable is the presence of endotoxin on the particle surface. The work of Daniels and colleagues[44] suggests that the biologic response to wear particles is a reflection of the response to surface endotoxin. However, numerous studies have elicited cytokine production by macrophages exposed to particles believed to be free of endotoxin.[45,46]

Corrosion age may have an important effect on the cellular response to metal particles with regard to the release of soluble metal species. Haynes and associates[47] demonstrated that the release of soluble metal from particles of comparable size and concentration to those found in hip joints (Co and Cr from CoCr particles and iron from 316L SS) was markedly reduced with time under physiologic conditions. Aged particles of both metals were much less toxic to monocytes than freshly produced particles and appeared to stimulate the release of greater amounts of IL-6 and PGE2.

Particle shape is another variable that is difficult to assess experimentally because of the confounding effects of surface area and size. Gelb and associates showed that small, irregularly shaped, mechanically produced particles of PMMA elicited a significantly greater inflammatory reaction than spherical particles of comparable size.[15] PE particles produced in vivo have characteristic shapes, primarily granular with varying proportions of fibrillar particles.[48] The mechanical irritation of the fibrillar PE particles may influence the cellular response in a manner comparable to the cellular response to chrysotile asbestos particles. The proteins binding to particles undoubtedly affect subsequent bioreactivity to the particle-protein complex, but the details have not yet been investigated.

Relevance

The production of PE wear particles from air-irradiated PE components satisfies two of the important criteria for adverse biologic reactions, size and concentration. Particle characterization studies have verified that PE particles are produced in the size range that makes them suitable for macrophage phagocytosis and activation. High wear rates of air-irradiated PE components can produce such an abundance of particles[49] that their concentration in the joint tissues can exceed the threshold for osteolysis in a matter of a few years. Fortunately the introduction of new types of wear-resistant PEs should significantly reduce the amount of wear particles and minimize the tissue response around total joint replacements. The addition of particles from sources other than the bearing surface (eg, from metallosis following component impingement) contributes to the particle burden in the tissues and may have direct toxic effects at high concentrations. Based on our current understanding of cell-particle interactions, reducing the number of wear particles produced by total joint replacements will reduce the adverse biologic responses to these particles.

Future Directions for Research

The biologic response to wear particles is a complex interaction involving multiple variables. The importance of any one stimulus may vary with the particular cell subtype or with the presence of microenvironmental stimuli. In vitro studies using molecular techniques have made a great contribution to our understanding of the factors that contribute to the cellular response to wear particles and the cellular mechanisms that drive them. However, in vitro studies using specific cell lines may have limited application to the in vivo situation. Extensive histologic studies of periprosthetic tissues demonstrate that similar biologic reactions occur with a wide range of materials. Willert and associates[50] concluded that the responses seen in vivo are not specific to any one material but result primarily from the amount of the particles and secondarily from the size of the particles. Further research on retrieved tissues and with cell models that incorporate multiple cell types exposed to particles of clinically relevant sizes, shapes, and concentrations may further elucidate the in vivo mechanisms.

References

1. Campbell P, Ma S, Yeom B, McKellop H, Schmalzried TP, Amstutz HC: Isolation of predominantly submicron-sized UHMWPE wear particles from periprosthetic tissues. *J Biomed Mater Res* 1995;29:127-131.
2. Doorn P, Campbell P, McKellop H, Benya P, Worrall J, Amstutz H: Characterization of metal wear particles from metal on metal total hip replacements. *Trans Soc Biomater* 1997;20:192.
3. Guttman D, Schmalzried TP, Jasty M, Harris WH: Light microscopic identification of submicron polyethylene wear debris. *J Appl Biomat* 1993;4:303-307.
4. Margevicius KJ, Bauer TW, McMahon JT, Brown SA, Merritt K: Isolation and characterization of debris in membranes around total joint prostheses. *J Bone Joint Surg Am* 1994;76:1664-1675.
5. Campbell P, McKellop H, Park SH, Dorr L: The clinical wear performance of modern generation metal-on-metal THRs: An implant retrieval study. *J Arthroplasty*, in press.
6. Goldring SR, Schiller AL, Roelke M, Rourke CM, O'Neill DA, Harris WH: The synovial-like membrane at the bone cement interface in loose total hip replacements and its proposed role in bone lysis. *J Bone Joint Surg Am* 1983;65:575-584.
7. Goodman SB, Huie P, Song Y, Schurman D, Maloney W, Woolson S, Sibley R: Cellular profile and cytokine production at prosthetic interfaces. *J Bone Joint Surg Br* 1998;80:531-539.
8. Jiranek WA, Machado M, Jasty M, et al: Production of cytokines around loosened cemented acetabular components: Analysis with immunohistochemical techniques and in situ hybridization. *J Bone Joint Surg Am* 1993;75:863-879.

9. Santavirta S, Sorsa T, Konttinen Y, Saari H, Eskola A, Eisen A: Role of mesenchymal collagenase in the loosening of total hip prosthesis. *Clin Orthop* 1993;290:206-215.

10. Shanbhag AS, Jacobs JJ, Black J, Galante JO, Glant TT: Cellular mediators secreted by interfacial membranes obtained at revision total hip arthroplasty. *J Arthroplasty* 1995;10:498-506.

11. Green TR, Fisher J, Stone M, Wroblewski BM, Ingham E: Polyethylene particles of a "critical size" are necessary for the induction of cytokines by macrophages in vitro. *Biomaterials* 1998;19:2297-2302.

12. North RJ: The relative importance of blood monocytes and fixed macrophages to the expression of cell-mediated immunity to infection. *J Exp Med* 1970;132:521-534.

13. Shanbhag AS, Jacobs JJ, Black J, Galante JO, Glant TT: Macrophage/particle interactions: Effect of size, composition and surface area. *J Biomed Mater Res* 1994;28:81-90.

14. Tabata Y, Ikada Y: Effect of the size and surface charge of polymer microspheres on their phagocytosis by macrophage. *Biomaterials* 1988;9:356-362.

15. Gelb H, Schumacher HR, Cuckler J, Baker DG: In vivo inflammatory response to polymethylmethacrylate particulate debris: Effect of size, morphology, and surface area. *J Orthop Res* 1994;12:83-92.

16. Horowitz SM, Doty SB, Lane JM, Burstein AH: Studies of the mechanism by which the mechanical failure of polymethylmethacrylate leads to bone resorption. *J Bone Joint Surg Am* 1993;75:802-813.

17. Doorn PF, Campbell PA, Worrall J, Benya PD, McKellop HA, Amstutz HC: Metal wear particle characterization from metal on metal total hip replacements: Transmission electron microscopy study of periprosthetic tissues and isolated particles. *J Biomed Mater Res* 1998;42:103-111.

18. Nakashima Y, Sun DH, Trindade MC, et al: Signaling pathways for tumor necrosis factor-alpha and interleukin-6 expression in human macrophages exposed to titanium-alloy particulate debris in vitro. *J Bone Joint Surg Am* 1999;81:603-615.

19. Rakshit D, Nestor B, Telford W, Salmon J, Sculco T: The role of cell surface receptors in particulate debris uptake by murine macrophages. *Trans Orthop Res Soc* 1998;23:121.

20. Sun JS, Liu HC, Chang WH, Li J, Lin FH, Tai HC: Influence of hydroxyapatite particle size on bone cell activities: An in vitro study. *J Biomed Mater Res* 1998;39:390-397.

21. Sochart DH: Relationship of acetabular wear to osteolysis and loosening in total hip arthroplasty. *Clin Orthop* 1999;363:135-150.

22. Dowd JE, Sychterz CJ, Young AM, Engh CA: Characterization of long-term femoral-head-penetration rates: Association with and prediction of osteolysis. *J Bone Joint Surg Am* 2000;82:1102-1107.

23. Kobayashi A, Freeman MA, Bonfield W, et al: Number of polyethylene particles and osteolysis in total joint replacements: A quantitative study using a tissue-digestion method. *J Bone Joint Surg Br* 1997;79:844-848.

24. Rader CP, Sterner T, Jakob F, Schutze N, Eulert J: Cytokine response of human macrophage-like cells after contact with polyethylene and pure titanium particles. *J Arthroplasty* 1999;14:840-848.

25. Gonzalez O, Smith RL, Goodman SB: Effect of size, concentration, surface area, and volume of polymethylmethacrylate particles on human macrophages in vitro. *J Biomed Mater Res* 1996;30:463-473.

26. Catelas I, Huk OL, Petit A, Zukor DJ, Marchand R, Yahia L: Flow cytometric analysis of macrophage response to ceramic and polyethylene particles: Effects of size, concentration, and composition. *J Biomed Mater Res* 1998;41:600-607.

27. Catelas I, Petit A, Zukor DJ, Marchand R, Yahia L, Huk OL: Induction of macrophage apoptosis by ceramic and polyethylene particles in vitro. *Biomaterials* 1999;20:625-630.

28. Rae T: The toxicity of metals used in orthopaedic prostheses: An experimental study using cultured human synovial fibroblasts. *J Bone Joint Surg Br* 1981;63:435-440.

29. Haynes DR, Rogers SD, Hay S, Pearcy MJ, Howie DW: The differences in toxicity and release of bone-resorbing mediators induced by titanium and cobalt-chromium-alloy wear particles. *J Bone Joint Surg Am* 1993;75:825-834.

30. Maloney WJ, Castro F, Schurman DJ, Smith RL: Effects of metallic debris on adult bovine articular chondrocyte metabolism in vitro. *J Applied Biomat* 1994;5:109-115.

31. Allen MJ, Myer BJ, Millett PJ, Rushton N: The effects of particulate cobalt, chromium and cobalt-chromium alloy on human osteoblast-like cells in vitro. *J Bone Joint Surg Br* 1997;79:475-482.

32. Evans EJ, Thomas IT: The in vitro toxicity of cobalt-chrome-molybdenum alloy and its constituent metals. *Biomaterials* 1986;7:25-29.

33. Haynes DR, Boyle SJ, Rogers SD, Howie DW, Vernon-Roberts B: Variation in cytokines induced by particles from different prosthetic materials. *Clin Orthop* 1998;352:223-230.

34. Horowitz SM, Luchetti WT, Gonzales JB, Ritchie CK: The effects of cobalt chromium upon macrophages. *J Biomed Mater Res* 1998;41:468-473.

35. Rogers SD, Howie DW, Graves SE, Pearcy MJ, Haynes DR: In vitro human monocyte response to wear particles of titanium alloy containing vanadium or niobium. *J Bone Joint Surg Br* 1997;79:311-315.

36. Wang JY, Wicklund BH, Gustillo RB, Tsukayama DT: Prosthetic metals impair murine immune response and cytokine release in vivo and in vitro. *J Orthop Res* 1997;15:688-699.

37. Shanbhag AS, Dowd JE, Jacobs JJ, et al: Biological response to particulate debris: In vitro and in vivo studies. *Cells and Materials* 1997;7:175-182.

38. Shanbhag AS, Jacobs JJ, Black J, Galante JO, Glant TT: Effects of particles on fibroblast proliferation and bone resorption in vitro. *Clin Orthop* 1997;342:205-217.

39. Lohmann CH, Schwartz Z, Koster G, et al: Phagocytosis of wear debris by osteoblasts affects differentiation and local factor production in a manner dependent on particle composition. *Biomaterials* 2000;21:551-561.

40. Haynes DR, Potter AE, Atkins GJ, et al: Metal particles stimulate expression of regulators of osteoclast development including osteoclast differentiation factor (RANKL/TRANCE), osteoprotegerin and macrophage colony stimulating factor. *Trans Orthop Res Soc* 1999;24;244.

41. Boynton EL, Waddell J, Meek E, Labow RS, Edwards V, Santerre JP: The effect of polyethylene particle chemistry on human monocyte-macrophage function in vitro. *J Biomed Mater Res* 2000;52:239-245.

42. Lazarus MD, Cuckler JM, Mitchell J, et al: Biocompatibility of PMMA with and without BaS in the rat subcutaneous air pouch model, in St John K (ed): *Particulate Debris From Medical Implants.* Philadelphia, PA, American Society for Testing and Materials, 1991.

43. Ingham E, Green TR, Stone MH, Kowalski R, Watkins N, Fisher J: Production of TNF-alpha and bone resorbing activity by macrophages in response to different types of bone cement particles. *Biomaterials* 2000;21:1005-1013.

44. Daniels AU, Barnes FH, Charlebois SJ, Smith RA: Macrophage cytokine response to particles and lipopolysaccharide in vitro. *J Biomed Mater Res* 2000;49:469-478.

45. Dean DD, Schwartz Z, Liu Y, et al: The effect of ultra-high molecular weight polyethylene wear debris on MG63 osteosarcoma cells in vitro. *J Bone Joint Surg Am* 1999;81:452-461.

46. Lazarus MD, Cuckler JM, Schumacher HR Jr, Ducheyne P, Baker DG: Comparison of the inflammatory response to particulate polymethylmethacrylate debris with and without barium sulfate. *J Orthop Res* 1994;12:532-541.

47. Haynes DR, Crotti TN, Haywood MR: Corrosion of and changes in biological effects of cobalt chrome alloy and 316L stainless steel prosthetic particles with age. *J Biomed Mater Res* 2000;49:167-175.

48. Campbell PA, Dorey F, McKellop HA, Amstutz HC: Morphological characterization of UHMWPE particles from total hips with alumina ceramic and cobalt femoral heads. *Trans Soc Biomater* 1995;18:308.

49. McKellop HA, Campbell P, Park SH, et al: The origin of submicron polyethylene wear debris in total hip arthroplasty. *Clin Orthop* 1995;311:3-20.

50. Willert HG, Buchhorn GH, Semlitsch M: Particle disease due to wear of metal alloys: Findings from retrieval studies, in Morrey BF (ed): *Biological, Material, Mechanical Considerations of Joint Replacement*. New York, NY, Raven Press, 1993, pp 129-146.

What is the role of endotoxin and fluid pressure in osteolysis?

The mechanisms by which wear particles affect osteolysis are not well understood. It is commonly accepted that ingestion of particles smaller than 10 μm in diameter causes activation of macrophages, resulting in release of proinflammatory cytokines and, ultimately, an increase in osteoclastic bone resorption. However, more recent studies indicate that the process is more complex, involving other cells in the periprosthetic environment as well as mechanical stimuli such as fluid pressure. These alternative mechanisms are the focus of this chapter.

Endotoxin

Endotoxin is the common name for components present in bacterial cell walls that induce the inflammatory response to bacteria. These complex molecules (including lipopolysaccharide (LPS), lipoteichoic acid (LTA), and peptidoglycan) and associated proteins induce responses similar to those caused by wear particles. The dimensions of wear particles are similar to those of bacteria. Although the process of phagocytosis may itself account for some of the similarities in response, the fact that the chemical entities LPS, LTA, and peptidoglycan also elicit these responses suggests that wear particles may contain endotoxins on their surface.

Endotoxins adhere to material surfaces, including wear particles. There is increasing evidence that endotoxins are still present even under aseptic conditions, and may augment the biologic responses induced by particle phagocytosis in macrophages.[1-5] Endotoxin can be derived either from subclinical infections on the implant surface or from the systemic circulation following transient bacteremia due to activities such as eating and tooth brushing.[6,7] It is very difficult to identify the sources of endotoxin because it is so prevalent and surfaces can be easily contaminated. Routine handling of specimens by technical staff can cause endotoxin contamination, and the common practice of isolating wear particles in the laboratory in the presence of serum virtually ensures it.

Nonorthopaedic literature emphasizes the importance of endotoxin in the immune response to particulates.[7] In dentistry studies, endotoxin has been shown to adsorb onto apatite surfaces, such as those found in the cementum coating the roots of teeth. All traces of endotoxin must be removed by careful and thorough root planing in order to prevent chronic inflammation, even after the bacterial infection is cured. It is becoming an increasingly important problem in peri-implantitis as well, because in this condition osseointegration of the surrounding alveolar bone with the dental implant is compromised and removal of endotoxin from the implant surface in situ is problematic.

When considering the role of endotoxin in the response of peri-implant cells to wear debris, one challenge is determining whether a biologic response is due to the material or to the endotoxin on the material. Ragab and associates suggest that some assays used to measure the presence of endotoxin are unable to detect endotoxin adherent on the surfaces of particles.[8] Thus some preparations of "endotoxin-free" particles may actually be contaminated. The concern about the contribution of endotoxin to cell response stimulated development of a method for isolating particles from formaldehyde-fixed tissues and measurement of endotoxin using the limulus assay used in clinical laboratories for detecting endotoxins.[9] Even with these precautions, however, some particles isolated from retrieved human tissue exhibited endotoxin contamination.

Endotoxins stimulate bone resorption by activating monocytes, as well as by mediating osteoclastogenesis.[10] More recently, endotoxins have been shown to have direct effects on osteoblasts, resulting in decreased proliferation and collagen synthesis, and inhibition of osteoblast differentiation.[11] To determine if endotoxin might also contribute to the response of osteoblasts to polyethylene particles, the response of human MG63 osteoblast-like cells to endotoxin-positive and endotoxin-negative particles from human retrieval tissues was compared.[9] Whether or not particles were contaminated with endotoxin, cell response was the same. Proliferation was increased and alkaline phosphatase-specific activity was decreased (Fig. 1, A). Moreover, production of transforming growth factor beta-1 (TGF-β1) was inhibited whereas production of prostaglandin E2 (PGE2) was stimulated (Fig. 1, B). In all instances the effects were dependent on the particle dose, suggesting that at least for these parameters, the dominant variable was the particle and not the endotoxin.

These studies indicate that endotoxins may play a significant role in the response of cells to wear particles. Endotoxins appear to be present on particles in vivo, even under aseptic conditions, and they are almost certainly present on particles prepared under nonsterile conditions in vitro. It is interesting to note that inclusion of antibiotics in bone cement may decrease the incidence of aseptic loosening by approximately 50%,[12] suggesting that microbial components can affect periprosthetic bone even in the absence of live organisms. Because many of the effects of endotoxin are similar to the effects ascribed to particles, it is important to understand the contributions of each to osteolysis and implant loosening. Both endotoxin and wear particles exert effects on monocytes and macrophages, causing a release of proinflammatory cytokines and activation of osteoclasts. Both have direct effects on osteoblasts, causing release of factors that activate osteoclasts. There are some key differences, however. Endotoxins either cause a decrease in proliferation of osteoblast-like cells or arrest proliferation, even in the presence

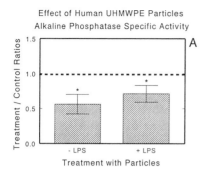
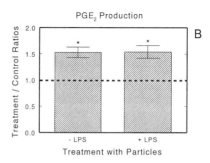

Figure 1 *Effect of endotoxin (LPS) contamination of polyethylene particles isolated from human tissue obtained at time of revision of failed hip prostheses. All particles were tested for presence of endotoxin using the limulus assay. Confluent cultures of MG63 human osteoblast-like cells were treated for 24 hours with 250×10^6 particles (mean diameter of 1 μm). A, Alkaline phosphatase specific activity was measured in cell layer lysates. B, PGE2 released into the conditioned medium was measured with a radioimmunoassay kit. Values are mean ± SEM of treatment/control ratios for 6 independent cultures. No statistical differences were found as a function of LPS presence or absence.*
$*P < 0.05$, treatment vs. control (=1)

of DNA synthesis.[11] In contrast, wear particles stimulate proliferation of osteoblast-like cells.[9,13,14] Thus osteoblasts may be less affected by endotoxin or particles than are macrophages. The relative contributions of endotoxin and wear particles to in vivo response are not yet known.

Fluid Pressure

Fluid pressure has also been implicated in osteolysis,[15-17] based on the anatomy of the bone-implant interface. If the bone is close to the surface of the implant, hydrostatic forces on the tissue are likely to have minimal consequences (Fig. 2, *A*). Unfortunately, this is often not the case. With the exception of initial bone formation on hydroxyapatite-coated surfaces, a fibrous connective tissue interface frequently exists between the implant and the bone surface. Even in the case of hydroxyapatite-coated surfaces, bone remodeling eventually resorbs the coating and the bone may no longer be directly adjacent to the implant. For cemented implants, bonding of bone to the implant surface does not occur.

While the bone-implant interface can provide mechanical interlock, it is exposed to intermittent fluctuating hydrostatic pressure (Fig. 2, *B*). Recent studies suggest that this pressure may contribute to osteolysis.[16] One mechanism by which this could occur is via the response of osteoblastic cells to changes in mechanical force. As shown in Figure 3, cells in the osteoblastic lineage are in contact with the bone-implant interface through canaliculi.

The major osteoblastic cell type in bone is the osteocyte, which is believed to be the cell responsible for transducing mechanical signals into biochemical signals. Osteocytes are in contact with each other and with osteoblasts and bone lining cells through gap junctions.[18] Because these cells are terminally

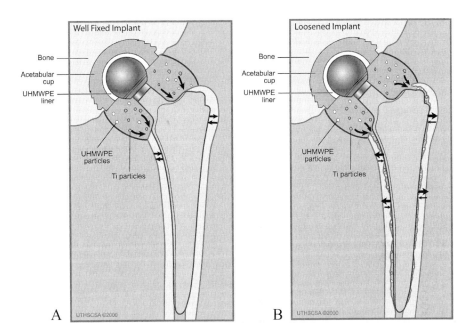

Figure 2 *Effect of fluid pressure on peri-implant osteolysis. Arrows denote primary direction of fluid pressure. **A,** Well fixed implant. The arrows indicate that the hydrostatic forces between the bone and implant are balanced and stable. **B,** Loosened implant. Arrows show that the hydrostatic forces are intermittent and fluctuating. (© 2000 UTHSCSA)*

differentiated osteoblasts, they do not synthesize and mineralize osteoid; however, they do regulate calcium and phosphate flux into and out of bone. These cells release factors (including nitric oxide [NO^{2-}], prostaglandins, interleukins, and TNF-α) that modulate osteoclasts and osteoblasts in the intercellular space. Thus exposure of these cells to increased hydrostatic force, or to increases in shear force through increased fluid flow in the canaliculi, can affect the rate and extent of bone resorption and subsequent bone formation.

It has become increasingly evident that osteoclast formation and activity are regulated by signals released from osteoblasts. Osteoprotogerin (OPG) is one such factor. In addition to producing RANK ligand (RANKL), osteoblasts also secrete a protein that binds to OPG. RANKL binds to its receptor on the osteoclast called RANK. Binding of OPG to RANKL prevents binding of RANKL to its receptors on osteoclasts, thereby limiting the upregulation of these bone resorbing cells. Childs and associates recently demonstrated that titanium wear debris-induced osteolysis is dependent on RANK signaling, as RANK knock-out mice are completely resistant and osteolysis in normal mice can be completely blocked by administration of a RANKL antagonist;[19] this finding provides another clue to the mechanisms by which these particles influence bone resorption.

Although these in vitro experiments are informative, actual contact of osteoblasts or osteocytes with wear particles in vivo has not been fully

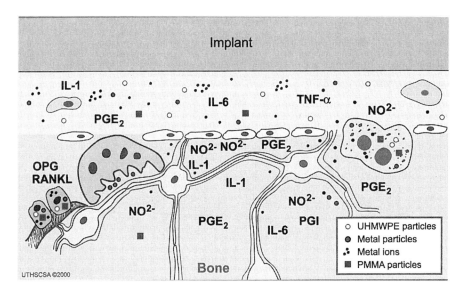

Figure 3 Communication of cells in the osteoblast lineage with the implant surface canaliculi. (© 2000 UTHSCSA)

assessed. Increased fluid flow into the tissue may not only cause a direct effect on the osteocyte by activating mechanoreceptors on the cells,[20] but it may facilitate the movement of wear particles into the tissue through the canaliculi. Because metallic wear debris particles have been observed at relatively large distances from an implant surface,[21] it is reasonable to anticipate that polyethylene and cement particles will move to these sites as well.

The question remains as to whether osteocytes are sensitive to wear particles. To test this, the responses of two osteoblastic cell lines to polyethylene particles were compared. OCT-1 cells are a well-differentiated osteoblast-like cell line derived from a transgenic mouse that was transfected with the SV40 large T-antigen driven by the osteocalcin promoter.[22] These cells synthesize osteoid and mineralize it in culture. MLO-Y4 cells were obtained from the same mouse and exhibit characteristics typical of osteocytes.[23] Both cell types respond to wear debris particles with an increase in nitric oxide production. The MLO-Y4 cells were more responsive to wear particles than the OCT-1 cells (Fig. 4).

These results indicate that fluid pressure may play two roles in the development of osteolysis. Fluid pressure may indirectly affect osteolysis by changing the hydrostatic forces, as well as the shear forces, on cells in the osteoblastic lineage and other cells, resulting in release of factors that activate osteoclast formation and activity. In addition, by increasing the movement of particulate debris into bone via the interface tissue and canaliculi, wear particles may exert direct effects on osteoblasts and osteocytes, causing release of factors that modulate bone resorption.

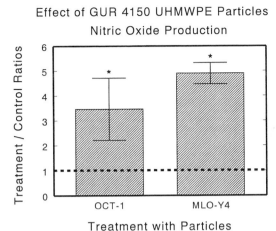

Figure 4 *Effect of GUR 4150 polyethylene particles on nitric oxide production by OCT-1 osteoblast-like cells and MLO-Y4 osteocyte-like cells. Confluent cultures of cells were treated with 250×10^6 particles for 24 hours. Nitric oxide production in the conditioned media was determined using Griess reagent normalized to the protein content of the cell layer. Values are treatment/control ratios for 6 cultures. No statistically significant differences were noted between cell types.*
$*P < 0.05$, treatment vs. control (=1)

Relevance

Osteolysis may result from mechanisms other than the phagocytosis of particles by monocytes and macrophages. Endotoxin may play a role in vivo even in the absence of acute infection. Similarly, fluctuating periprosthetic pressure and fluid flow during gait and other activities may result in increased shear force on osteocytes within the bone canaliculi. Factors released by these cells in response to mechanical stress may impact the rate and extent of osteoclastic bone resorption.

Future Directions for Research

Additional research should be directed to better understand the role of the osteocyte in response to wear debris; to better understand the role of fluid flow, hydrostatic pressure, and other mechanical forces in osteolysis; and to determine the relative contributions of particles and contaminating endotoxin in aseptic loosening and osteolysis.

References

1. Cho DR, Hong CY, Baran GR, Goldring SR: Adsorbed endotoxin mediates differential effects on particle-induced stimulation of cytokine and chemokine release. *Trans Orthop Res Soc* 1999;24:6.
2. Daniels AU, Barnes FH, Charlebois SJ, Smith RA: Macrophage cytokine response to particles and lipopolysaccharide in vitro. *J Biomed Mater Res* 2000;49:469-478.

3. Hitchins VM, Merritt K: Decontaminating particles exposed to bacterial endotoxin (LPS). *J Biomed Mater Res* 1999;46:434-437.

4. Akisue T, Bauer T, Farver C: The effect of particle wear debris on nuclear factor kappa B activation in differentiated THP-1 cells. *Trans Orthop Res Soc* 2000;25:594.

5. Aspenberg P: Adherent endotoxins are necessary for particle-induced bone resorption in a rat model. *Trans Orthop Res Soc* 2000;25:704.

6. Bi Y, Kaar SG, Ragab AA, Goldberg VM, Anderson JM, Greenfield EM: Adherent endotoxin on orthopaedic wear particles stimulates cytokine production and osteoclast differentiation: Mediation by toll-like receptor-4. *J Bone Miner Res*, in press.

7. Robinson FG, Knoernschild KL, Sterrett JD, Tompkins GR: Porphyromonas gingivalis endotoxin affinity for dental ceramics. *J Prosthet Dent* 1996;75:217-227.

8. Ragab AA, Van De Motter R, Lavish SA, et al: Measurement and removal of adherent endotoxin from titanium particles and implant surfaces. *J Orthop Res* 1999;17:803-809.

9. Dean DD, Schwartz Z, Liu Y, et al: The effect of ultrahigh molecular weight polyethylene wear debris on MG63 osteosarcoma cells in vitro. *J Bone Joint Surg Am* 1999;81:452-461.

10. Abu-Amer Y, Ross FP, Edwards, J, Teitelbaum SL: Lipopolysaccharide-stimulated osteoclastogenesis is mediated by tumor necrosis factor via its P55 receptor. *J Clin Invest* 1997;100:1557-1565.

11. Nair SP, Meghji S, Wilson M, Reddi K, White P, Henderson B: Bacterially induced bone destruction: Mechanisms and misconceptions. *Infect Immun* 1996;64:2371-2380.

12. Espehaug B, Engesaeter LB, Vollset SE, Havelin LI, Langeland N: Antibiotic prophylaxis in total hip arthroplasty: Review of 10,905 primary cemented total hip replacements reported to the Norwegian arthroplasty register, 1987 to 1995. *J Bone Joint Surg Br* 1997;79:590-595.

13. Dean DD, Schwartz Z, Blanchard CR, et al: Ultrahigh molecular weight polyethylene (UHMWPE) particles have direct effects on proliferation, differentiation, and local factor production of MG63 osteoblast-like cells. *J Orthop Res* 1999;17:9-17.

14. Lohmann CH, Schwartz Z, Köster G, et al: Phagocytosis of wear debris by osteoblasts affects differentiation and local factor production in a manner dependent on particle composition. *Biomaterials* 2000;21:551-561.

15. van der Vis H, Aspenberg P, de Kleine R, Tigchelaar W, van Noorden CJ: Short periods of oscillating fluid pressure directed at a titanium-bone interface in rabbits lead to bone lysis. *Acta Orthop Scand* 1998;69:5-10.

16. van der Vis H, Aspenberg P, Marti RK, Tigchelaar W, van Noorden CJ: Fluid pressure causes bone resorption in a rabbit model of prosthetic loosening. *Clin Orthop* 1998;350:201-208.

17. Aspenberg P, van der Vis H: Fluid pressure may cause periprosthetic osteolysis: Particles are not the only thing. *Acta Orthop Scand* 1998;69:1-4.

18. Shapiro F: Variable conformation of GAP junctions linking bone cells: A transmission electron microscopic study of linear, stacked linear, curvilinear, oval, and annular junctions. *Calcif Tissue Int* 1997;61:285-293.

19. Childs L, Paschalis E, Shigeyama Y, et al: Long term protection from wear debris-induced bone resorption and amelioration of established osteolysis by RANK:Fe. *J Bone Miner Res* 2000;15(suppl 1):S192.

20. Cheng B, Kato Y, Zhao S, Luo J, Sprague E, Bonewald LF, Jiang JX: Prostaglandin is essential for gap junction-mediated intercellular communication between osteocytes in response to mechanical strain (Abstract SU188). *J Bone Miner Res* 2000;15:S377.

21. Willert HG, Buchhorn GH, Göbel D, et al: Wear behavior and histopathology of classic cemented metal on metal hip endoprostheses. *Clin Orthop* 1996;329S:S160-S186.

22. Chen D, Chen H, Feng JQ, et al: Osteoblastic cell lines derived from a transgenic mouse containing the osteocalcin promoter driving the SV40 T-antigen. *Mol Cell Differentiation* 1995;3:193-212.

23. Kato Y, Windle JJ, Koop BA, Mundy GR, Bonewald LF: Establishment of an osteocyte-like cell line, MLO-Y4. *J Bone Miner Res* 1997;12:2014-2023.

What experimental approaches (tissue retrieval, in vivo, in vitro, etc) have been used to investigate the biologic effects of particles?

In investigating osteolysis and aseptic loosening around total joint replacements, the traditional approach has involved radiographic evaluation and analysis of retrieved tissues. Based on these analyses, experimental models have been developed to identify and confirm the etiologic agents.

When Charnley observed loosening and osteolysis in joint replacements associated with polytetrafluorethylene (PTFE) in the 1960s, he recognized a relationship between wear particles and granulomatous inflammation based on analysis of retrieved tissues and personal experimentation.[1-4] Histopathologic analysis of tissues retrieved during revision surgery indicated the presence of particulate material surrounded by foreign body giant cells forming "caseating granulomata and sterile pus." To confirm that particulate PTFE debris caused the granuloma, Charnley implanted finely divided PTFE subcutaneously in his own thigh and studied the increased size of the nodules, which he could palpate after 3 months.[2] Extensive histopathologic analyses of retrieved tissues also led Willert and Semlitsch to hypothesize that particulate wear debris leads to a macrophage response with the ability to disrupt the prosthetic attachment to bone and lead to implant loosening (Fig. 1).[5,6] Experimental studies have demonstrated that the particle-mediated biologic responses and mechanical factors play important roles in causing inflammation that leads to periprosthetic bone resorption.

Analyses of Clinical Material

Histopathologic analysis of clinical material (Fig. 2) is the primary technique used to investigate the pathogenesis of osteolysis and aseptic loosening. During these analyses, the cells, particulate debris, and biologic mediators present are identified and localized.

The clinical materials analyzed may be harvested during biopsies, revision surgery, or autopsies. Investigators have studied the synovial fluid, periprosthetic tissues, capsular tissues, and extra-articular tissues such as lymph nodes. During autopsies, the implant and surrounding soft tissues and

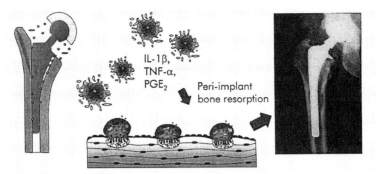

Figure 1 *Particle-mediated osteolysis. Schematic representation of the role of wear debris generated at the articular surfaces in stimulating macrophages to release bone-resorbing mediators that can in turn stimulate osteoclastic bone resorption, leading to osteolysis and aseptic loosening.*

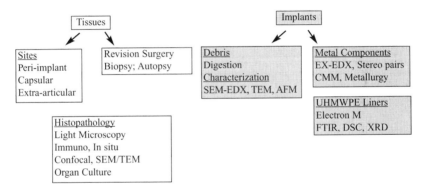

Figure 2 *Analysis of clinical materials. Different techniques used in characterizing tissues and implant materials from joint replacements.*

bone are harvested; regional and distant lymph nodes and organ tissue from the lungs, liver, and spleen also yield critical information.

The choice of technique to examine the tissues depends on the specific question that is posed. Light microscopy of hematoxylin and eosin-stained tissue sections is routine and provides valuable information on the general organization of the tissues and cells associated and involved in periprosthetic bone loss.[3,5,7-15] Using specific staining techniques as well as immunohistochemistry, the cell types involved in the chronic inflammatory process can be identified and the temporospatial relationship between tissues and cells can be established.[16-19] The role of macrophages and giant cells in internalizing or walling off wear debris was also demonstrated, as well as the association of the granuloma with the bony erosions. Histologic studies have also noted the absence of acute inflammatory cells in the interface tissues.[8,11,12,14,20]

Placing the retrieved tissues in organ culture allows identification of the cellular mediators that are released by the interface tissue, which in turn can provide insight into the microenvironment of the cells. Goldring and associates used organ culture techniques to demonstrate that periprosthetic tissues were capable of producing biologic mediators with the capacity to stimulate osteoclasts to resorb bone.[21,22] Other investigators have subsequently shown that the cellular activity within this membrane is capable of producing a variety of matrix-degrading enzymes, prostaglandins, and proinflammatory cytokines and chemokines that are capable of stimulating osteoclastic bone resorption and fibrous tissue formation.[10,22-26]

Similar findings have also been reported using enzymatic digestion to disperse cell populations and then analyzing the products released in culture. Although organ culture techniques are useful in identifying the inflammatory mediators involved in the osteolytic process, in situ hybridization techniques are essential to identify the specific cell types and correlate particle internalization and gene expression for specific mediators.[17,18,27,28] Confocal microscopy and electron microscopy have also been used to establish the intracellular spatial organization of the particles in the context of the organelles and lysosomes.[29]

Retrieved tissues also provide a unique opportunity to study the particulate wear debris present in osteolytic tissues. Various techniques have been used to digest the tissues and extract the particulate debris present. The physicochemical characteristics of these retrieved particles are then determined with scanning electron microscopy, transmission electron microscopy, and atomic force microscopy.[30-38] These findings have provided critical information in elucidating biologic reactivity and have enabled the development of clinically relevant experimental models.

Material and metallurgic analyses of the retrieved implants themselves also provide such critical information as scratches on the femoral head, abrasion of the metal stem, delamination of the fiber-metal porous pad and metal beads, and fracture of wires and cables used during surgery. Techniques used include stereomicroscopy, electron microscopy, and metallurgic analyses to study metal grain structure, distribution, and corrosion. Analyses of retrieved acetabular liners have yielded information on consolidation of the polyethylene and sterilization and shelf storage of the liners. The techniques used in these cases include Fourier transform infrared spectroscopy, differential scanning calorimetry to identify the chemical species, and x-ray diffraction (wide and small angles) to study crystallinity.

Experimental Models

Retrieved clinical material provides valuable information to elucidate the events occurring at the bone-implant interface. However, in vivo and in vitro models are required to demonstrate that the addition of debris results in the formation of granulomatous tissues and implant loosening and confirm that particulate wear debris is the etiologic agent in osteolysis (Fig. 3).

Early studies using rabbits, rats, and guinea pigs were performed mainly to investigate the biocompatibility of bulk and particulate materials. Materials of various sizes, shapes, and compositions were implanted in different subcutaneous, intramuscular, or intra-articular locations. In one of the earliest studies (1959), Cohen[39] studied the subcutaneous tissue response in rabbits to different forms and sizes of commonly used CoCr alloy and stain-

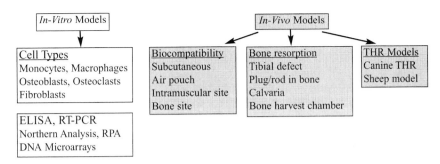

Figure 3 *Experimental models. Schematic representation of in vitro and in vivo models used to investigate the mechanisms leading to periprosthetic bone loss.*

less steel particles. He reported that larger particles (5 to 40 µm) provoked very little inflammatory response that took the form of a fibrous tissue encapsulation. Finer particles provoked an inflammatory reaction associated with tissue necrosis, acute inflammation, and extensive fibrosis.[39] Studies have confirmed this host response, which forms the basis for concern over fine wear debris around total joint arthroplasties.[13,40-42] More recently, a subcutaneous rat air-pouch model was used to demonstrate that smaller irregularly shaped particles elicited significantly greater levels of inflammatory mediators than did smoother particles.[43]

More sophisticated models have demonstrated the inflammatory response of particulate wear debris in a bone location. With a rabbit tibial defect model, Goodman and associates demonstrated that introduction of particulate debris can lead to development of a florid fibrohistiocytic and giant cell reaction, similar to that surrounding loose implants.[40,44-46] Bulk forms of the same materials were surrounded only by a thin fibrous membrane.[25,45,47] Howie and associates initially modeled the implantation of a bulk implant in an intramedullary cavity using a polymethylmethacrylate (PMMA) plug in the femoral condyle of a rat knee.[48] Repeated injections of polyethylene particles in the knee resulted in resorption of the bone surrounding the plugs. Osteolysis surrounding the cement plug was associated with fibrous tissue containing macrophages, giant cells, and polyethylene particles.[48] Murine or rat calvaria provide an invaluable site for studying the bone resorption ability of foreign-body granuloma.[49]

In other models, partial or total hip replacements have been performed in a larger animal, with the introduction of particulate wear debris. Spector and associates[50] and Turner and colleagues[51] modeled a loose femoral stem in canines by overreaming the femoral canal and introduced PMMA or bone particles. Both groups observed a characteristic granulomatous reaction and areas of endosteal bone loss. Two teams developed canine models for press-fit uncemented total hip replacements.[52,53] A fiber metal-coated porous TiAlV femoral component articulated against a polyethylene acetabular liner with a fiber mesh-backed metal shell. Different types of particulate wear debris were introduced intraoperatively at the bone-implant interface. Recessing the proximal implant 2 mm created a gap at the proximal bone-implant interface and facilitated introduction of wear debris. After 12 weeks, a granulomatous tissue was observed at the bone-implant interface that histologically resembled the tissues from around clinically loosened implants. Furthermore, when these tissues were placed in organ cultures, they released significant levels of proinflammatory substances such as IL-1, PGE2, and collagenase and gelatinase.[52] By extending the postoperative period to 24 weeks, radiographic evidence of periprosthetic bone loss was observed. The pattern of bone loss and associated periosteal reaction was similar to loosening and osteolysis in humans.[53]

Although analysis of in vivo models provides a valuable contribution to an understanding of the events occurring at the bone-implant interface, data are confounded by factors such as particles generated in situ, paracrine effects, and the contributions of remote organs. In vitro cell culture models permit the introduction of defined wear debris species into specific cell types. This allows the evaluation of particle cytotoxicity, the effect of pure and mixed particle species, size and dose response of particles, and the type and quantity of the pro- and anti-inflammatory mediators released. Additionally, these in vitro systems provide an experimental environment to investigate specific molecular responses to debris exposure. In developing cell culture systems, investigators have studied the cell types that are routinely identified in the periprosthetic tissues. The interaction of tissue macrophages with predominantly phagocytosable debris appears to be a key event in the initiation of the inflammatory cascade, the formation of the periprosthetic granulomatous tissue, and associated bone resorption. Macrophages cultured with particles in vitro are thus the most common means for studying particle-cellular interactions at the bone-implant interface.

Transformed or immortalized macrophage cell lines such as $P388D_1$, J774, IC-21, or RAW 267 have been used extensively to understand macrophage-particle interactions.[27,54-59] Their advantage lies in the ability to harvest large numbers of nearly identical cells in order to conduct studies with substantially reduced intragroup variability. Other investigators have used murine peritoneal macrophages,[60,61] peripheral blood monocytes from human volunteers,[62] or mononuclear cells isolated from interfacial membranes.[63] In addition to macrophages, investigators have also studied the interaction of particles with fibroblasts and osteoblasts.[64-68] These cells have important interactions with macrophages, either directly through released mediators or by their influence on bone resorption and formation.

Early studies on cell-particle interactions focused on the toxicities of different particle compositions. The most common techniques were DNA synthesis using a radioactive isotope (tritiated thymidine) or colorimetric assays (MTT).[57,60,69] Damage to cell membranes can also be assessed by quantifying extracellular lactate dehydrogenase release.[54,55] The predominant manner of defining cell-particle interaction involved determining the profile of different mediators released.[27,41,55,56,61,70] The typical technique used to identify and quantify the levels of translated protein is the enzyme-linked immunosorbent assay.[41,62,71] Reverse transcriptase-polymerase chain reaction is used to identify the messenger RNA for transcription products, and Northern analysis is used to quantify them.[67,68] The molecular pathways have been elucidated with techniques such as gel-shift assays[72] and DNA microarrays to identify thousands of genes expressed after cell-particle interaction.[73]

Relevance

The use of cell and organ culture, in vitro and in vivo models, and analysis of retrieved tissues and implants will aid in the understanding of the biologic events associated with loosening and osteolysis. Strategies can then be developed to mitigate adverse biologic events that jeopardize implant longevity.

Future Directions for Research

There is a significant need for standardized particles for in vivo and in vitro experiments to facilitate comparison of results from different models. With the sequencing of the human genome, new technologies are being developed to enhance our understanding of the pathogenesis of aseptic loosening and osteolysis. Messenger RNA extracted from retrieved tissues can be hybridized to DNA microarrays and the profile of gene expression can be determined. This information can be correlated with patient variables, progression of disease, therapeutic regimens, and types of materials used and the molecular pathways involved in periprosthetic bone loss can then be identified. Because these techniques will generate a large volume of data, however, improved and substantiated methods of data analysis must be developed, such as shared implant tissue libraries and web accessible data sets. The significant effort in developing gene transfer therapies could enable gene insertion to prevent osteolysis.

References

1. Charnley J: Anchorage of the femoral head prosthesis to the shaft of the femur. *J Bone Joint Surg Br* 1960;42:28-30.
2. Charnley J: Letter: Tissue reactions to polytetrafluorethylene. *Lancet* 1963:1379.
3. Charnley J: Tissue reaction to implanted plastics, in Charnley J (ed): *Acrylic Cement in Orthopedic Surgery*. Edinburgh, Scotland, E & S Livingstone, 1970, pp 1-9.
4. Charnley J, Follacci FM, Hammond BT: The long-term reaction of bone to self-curing acrylic cement. *J Bone Joint Surg Br* 1968;50:822-829.
5. Willert HG, Semlitsch M: Tissue reactions to plastic and metallic wear products of joint endoprostheses, in Gschwend N, Debrunner HU (eds): *Total Hip Prosthesis*. Baltimore, MD, Williams and Wilkins, 1976, pp 205-239.
6. Willert HG, Semlitsch M: Reactions of the articular capsule to wear products of artificial joint prostheses. *J Biomed Mater Res* 1977;11:157-164.

7. Forest M, Carlioz A, Vacher Lavenu MC, et al: Histological patterns of bone and articular tissues after orthopaedic reconstructive surgery (artificial joint implants). *Path Res Pract* 1991;187:963-977.

8. Harris WH, Schiller AL, Scholler J-M, Freiberg RA, Scott R: Extensive localized bone resorption in the femur following total hip replacement. *J Bone Joint Surg Am* 1976;58:612-618.

9. Howie DW: Tissue response in relation to type of wear particles around failed hip arthroplasties. *J Arthroplasty* 1990;5:337-348.

10. Kim KJ, Rubash HE, Wilson SC, D'Antonio JA, McClain EJ: A histological and bio-chemical comparison of the interface tissues in cementless and cemented hip prosthe-sis. *Clin Orthop* 1993;287:142-152.

11. Mirra JM, Amstutz HC, Matos M, Gold R: The pathology of the joint tissues and its clinical relevance in prosthesis failure. *Clin Orthop* 1976;117:221-240.

12. Mirra JM, Marder RA, Amstutz HC: The pathology of failed total joint arthroplasty. *Clin Orthop* 1982;170:175-183.

13. Sullivan PM, MacKenzie JR, Callaghan JJ, Johnston RC: Total hip arthroplasty with cement in patients who are less than fifty years old. *J Bone Joint Surg Am* 1994;76:863-869.

14. Vernon-Roberts B, Freeman MAR: Morphological and analytical studies of the tis-sues adjacent to joint prostheses: Investigations into the causes of loosening prosthe-ses, in Schaldach M, Hohmann D (eds): *Advances in Artificial Hip and Knee Joint Technology.* Berlin, Germany, Springer-Verlag, 1976, pp 148-186.

15. Willert HG, Ludwig J, Semlitsch M: Reaction of bone to methacrylate after hip arthroplasty. *J Bone Joint Surg Am 1974;56:1368-1382.*

16. Chiba J, Schwendeman LJ, Booth RE Jr, Crossett LS, Rubash HE: A biochemical, histologic and immunologic analysis of membranes obtained from failed cemented and cementless total knee arthroplasty. *Clin Orthop* 1994;299:114-124.

17. Goodman SB, Knoblich G, O'Connor M, Song Y, Huie P, Sibley R: Heterogeneity in cellular and cytokine profiles from multiple samples of tissue surrounding revised hip prostheses. *J Biomed Mater Res* 1996;31:421-428.

18. Jiranek WA, Machado M, Jasty MJ, et al: Production of cytokines around loosened cemented acetabular components: Analysis with immunohistochemical techniques and in situ hybridization. *J Bone Joint Surg Am* 1993;75:863-879.

19. Macaulay W, Watkins S, Turner D, Rubash HE, Evans CH: Immunohistochemical identification of nitric oxide synthase in interfacial membranes from loosened hip prostheses. *Trans Orthop Res Soc* 1995;20:221.

20. Vernon-Roberts B, Freeman MAR: The tissue response to total joint replacement prostheses, in Swanson SAV, Freeman MAR (eds): *The Scientific Basis of Joint Replacement.* New York, NY, John Wiley & Sons, 1977, pp 86-129.

21. Goldring SR, Jasty MJ, Roelke MS, Rourke CM, Bringhurst FR, Harris WH: Formation of a synovial-like membrane at the bone-cement interface: Its role in bone resorption and implant loosening after total hip replacement. *Arthritis Rheum* 1986;29:836-841.

22. Goldring SR, Schiller AL, Roelke M, Rourke CM, O'Neill DA, Harris WH: The syn-ovial-like membrane at the bone-cement interface in loose total hip replacements and its proposed role in bone lysis. *J Bone Joint Surg Am* 1983;65:575-584.

23. Chiba J, Rubash HE, Kim KJ, Iwaki Y: The characterization of cytokines in the inter-face tisssue obtained from failed cementless total hip arthroplasty with and without femoral osteolysis. *Clin Orthop* 1994;300:304-312.

24. Dorr LD, Bloebaum R, Emmanual J, Meldrum R: Histologic, biochemical and ion analysis of tissue and fluids retrieved during total hip arthroplasty. *Clin Orthop* 1990;261:82-95.

25. Goodman SB, Chin RC: Prostaglandin E2 levels in the membrane surrounding bulk and particulate polymethylmethacrylate in the rabbit tibia. *Clin Orthop* 1990;257:305-309.

26. Shanbhag AS, Jacobs JJ, Black J, Galante JO, Glant TT: Cellular mediators secreted by interfacial membranes obtained at revision total hip arthroplasties. *J Arthroplasty* 1995;10:498-506.

27. Glant TT, Jacobs JJ, Molnár G, Shanbhag AS, Valyon M, Galante JO: Bone resorption activity of particulate-stimulated macrophages. *J Bone Miner Res* 1993;8:1071-1079.

28. Kato K, Yokoi T, Takano N, et al: Detection of in situ hybridization and phenotypic characterization of cells expressing IL-6 mRNA in human stimulated blood. J *Immunol* 1990;144:1317-1322.

29. Benz EB, Federman M, Godleski JJ, et al: Ultrastructure of cells that have phagocytosed polyethylene particles in peri-implant tissue from revision arthroplasty. *Trans Orthop Res Soc* 1994;19:200.

30. Campbell P, Ma S, Schmalzried TP, Amstutz HC: Tissue digestion for wear debris particle isolation. *J Biomed Mater Res* 1994;28:523-526.

31. Campbell P, Ma S, Yeom B, McKellop HA, Schmalzried TP, Amstutz HC: Isolation of predominantly submicron-sized UHMWPE wear particles from periprosthetic tissues. *J Biomed Mater Res* 1995;29:127-131.

32. Klimkiewicz JJ, Iannotti JP, Rubash HE, Shanbhag AS: Aseptic loosening of the humeral component in total shoulder arthroplasty. *J Shoulder Elbow Surg* 1998;7:422-426.

33. Kossovsky N, Liao K, Millett D, et al: Periprosthetic chronic inflammation characterized through the measurement of superoxide anion production by synovial-derived macrophages. *Clin Orthop* 1991;263:263-271.

34. Maloney WJ, Smith RL, Schmalzried TP, Chiba J, Huene D, Rubash HE: Isolation and characterization of wear particles generated in patients who have had failure of a hip arthroplasty without cement. *J Bone Joint Surg Am* 1995;77:1301-1310.

35. Margevicius KJ, Bauer TW, McMahon JT, Brown SA, Merritt K: Isolation and characterization of debris in membranes around total joint prostheses. *J Bone Joint Surg Am* 1994;76:1664-1675.

36. Schmalzried TP, Campbell P, Schmitt AK, Brown IC, Amstutz HC: Shapes and dimensional characteristics of polyethylene wear particles generated in vivo by total knee replacements compared to total hip replacements. *J Biomed Mater Res* 1997;38:203-210.

37. Shanbhag AS, Jacobs JJ, Glant TT, Gilbert JL, Black J, Galante JO: Composition and morphology of wear debris in failed uncemented total hip replacement arthroplasty. *J Bone Joint Surg Br* 1994;76:60-67.

38. Wolfarth DL, Hahn DW, Bushar G, Parks NL: Separation and characterization of polyethylene wear debris from synovial fluid and tissue samples of revised knee replacements. *J Biomed Mater Res* 1997;34:57-61.

39. Cohen J: Assay of foreign-body reaction. *J Bone Joint Surg Am* 1959;41:152-166.

40. Goodman SB, Fornasier VL, Lee J, Kei J: The histological effects of the implantation of different sizes of polyethylene particles in the rabbit tibia. *J Biomed Mater Res* 1990;24:517-524.

41. Horowitz SM, Doty SB, Lane JM, Burstein AH: Studies of the mechanism by which the mechanical failure of polymethylmethacrylate leads to bone resorption. *J Bone Joint Surg Am* 1993;75:802-813.

42. Escalas F, Galante JO, Rostoker W, Coogan P: Biocompatibility of materials for total joint replacement. *J Biomed Mater Res* 1976;10:175-195.
43. Sporn MB, Roberts AB, Wakefield LM, Assoian RK: Transforming growth factor-b: Biological function and chemical structure. *Science* 1986;233:532-534.
44. Goodman SB, Chin RC, Chiou SS, Lee JS: Prostaglandin E2 synthesis by the tissue surrounding ultrahigh molecular weight polyethylene in different physical forms, in St John KR (ed): *Particulate Debris from Medical Implants: Mechanisms of Formation and Biological Consequences.* Philadelphia, PA, American Society for Testing and Materials, 1992, pp 111-117.
45. Goodman SB, Fornasier VL, Kei J: The effects of bulk versus particulate polymethyl-methacrylate on bone. *Clin Orthop* 1988;232:255-262.
46. Goodman SB, Fornasier VL, Kei J: The effects of bulk versus particulate ultra-high-molecular-weight polyethylene on bone. *J Arthroplasty* 1988;3(suppl):S41-S46.
47. Miller KM, Anderson JM: Human monocyte/macrophage activation and interleukin 1 generation by biomedical polymers. *J Biomed Mater Res* 1988;22:713-731.
48. Howie DW, Vernon-Roberts B, Oakeshott R, Manthey B: A rat model of resorption of bone at the cement-bone interface in the presence of polyethylene wear particles. *J Bone Joint Surg Am* 1988;70:257-263.
49. Merkel KD, Erdmann JM, McHugh KP, Abu-Amer Y, Ross FP, Teitelbaum SL: Tumor necrosis factor-alpha mediates orthopedic implant osteolysis. *Am J Path* 1999;154:203-210.
50. Spector M, Shortkroff S, Hsu H-P, Lane N, Sledge CB, Thornhill TS: Tissue changes around loose prostheses: A canine model to investigate the effects of an antiinflamma-tory agent. *Clin Orthop* 1990;261:140-152.
51. Turner TM, Urban RM, Sumner DR, Galante JO: Revision, without cement, of asepti-cally loose, cemented total hip prostheses. *J Bone Joint Surg Am* 1993;75:845-862.
52. Dowd JE, Schwendeman LJ, Macaulay W, et al: Aseptic loosening in uncemented total hip arthroplasty in a canine model. *Clin Orthop* 1995;319:106-121.
53. Hasselman CT, Shanbhag AS, Kovach C, Marinelli R, Rubash HE: Osteolysis and aseptic loosening in a canine uncemented total hip arthroplasty (THA) model. *Trans Orthop Res Soc* 1997;22:22.
54. Horowitz SM, Frondoza CG, Lennox DW: Effects of polymethylmethacrylate expo-sure upon macrophages. *J Orthop Res* 1988;6:827-832.
55. Horowitz SM, Gautsch TL, Frondoza CG, Riley L Jr: Macrophage exposure to poly-methyl methacrylate leads to mediator release and injury. *J Orthop Res* 1991;9:406-413.
56. Macaulay W, Shanbhag AS, Marinelli R, Stefanovic-Racic M, Rubash HE, Woo SL-Y: Nitric oxide release from murine macrophages when stimulated with particulate debris. *Trans Soc Biomater* 1995;18:307.
57. Rae T: A study on the effects of particulate metals of orthopaedic interest on murine macrophages in vitro. *J Bone Joint Surg Br* 1975;57:444-450.
58. Schindler R, Mancilla J, Endres S, Ghorbani R, Clarke SC, Dinarello CA: Correlations and interactions in the production of interleukin-6 (IL-6), IL-1 and tumor necrosis factor (TNF) in human blood mononuclear cells: IL-6 suppresses IL-1 and TNF. *Blood* 1990;75:40-47.
59. Shanbhag AS, Jacobs JJ, Black J, Galante JO, Glant TT: Macrophage/particle interac-tions: Effect of size, composition and surface area. *J Biomed Mater Res* 1994;28:81-90.
60. Garrett R, Wilksch J, Vernon-Roberts B: Effects of cobalt-chrome alloy wear particles on the morphology, viability and phagocytic activity of murine macrophages in vitro. *Aust J Exp Biol Med Sci* 1983;61:355-369.

61. Glant TT, Jacobs JJ: Response of three murine macrophage populations to particulate debris: Bone resorption in organ cultures. *J Orthop Res* 1994;12:720-731.

62. Shanbhag AS, Jacobs JJ, Black J, Galante JO, Glant TT: Human monocyte response to particulate biomaterials generated in-vivo and in-vitro. *J Orthop Res* 1995;13:792-801.

63. Lee SH, Brennan FR, Jacobs JJ, Urban RM, Ragasa DR, Glant TT: Human monocyte/macrophage response to particulate wear debris: Sensitization of macrophages to cobalt-chromium corrosion products and titanium in patients with total joint replacements. *J Orthop Res* 1997;15:40-49.

64. Dean DD, Schwartz DE, Liu Y, et al: The effect of ultra-high molecular weight polyethylene wear debris on MG 63 osteosarcoma cells in vitro. *J Bone Joint Surg Am* 1999;81:452-461.

65. Lind M, Trindade MCD, Yaszay B, Goodman SB, Smith RL: Effects of particulate debris on macrophage-dependent fibroblast stimulation in coculture. *J Bone Joint Surg Br* 1998;80:924-930.

66. Maloney WJ, Smith RL, Castro F, Schurman DJ: Fibroblast response to metallic debris in vitro: Enzyme induction, cell proliferation, and toxicity. *J Bone Joint Surg Am* 1993;75:835-844.

67. Yao J, Cs-Szabo G, Jacobs JJ, Kuettner KE, Glant TT: Suppression of osteoblast function by titanium particles. *J. Bone Joint Surg Am* 1997;79:107-112.

68. Yao J, Glant TT, Lark MW, et al: The potential role of fibroblasts in periprosthetic osteolysis: Fibroblast response to titanium particles. *J Bone Miner Res* 1995;10:1417-1427.

69. Rae T: The biological response to titanium and titanium-aluminum-vanadium alloy particles: I. Tissue culture studies. *Biomaterials* 1986;7:30-36.

70. Gelb H, Schumacher HR, Cuckler J, Baker DG: In vivo inflammatory response to polymethylmethacrylate particulate debris: Effect of size, morphology, and surface area. *J Orthop Res* 1994;12:83-92.

71. Haynes DR, Rogers SD, Hay S, App B, Pearcy MJ, Howie DW: The differences in toxicity and release of bone-resorbing mediators induced by titanium and cobalt-chromium-alloy wear particles. *J Bone Joint Surg Am* 1993;75:825-834.

72. Nakashima Y, Sun D-H, Trindade MCD, Maloney WJ, Goodman SB, Schurman DJ, Smith RL: Signaling pathways for tumor necrosis factor-α and interleukin-6 expression in human macrophages exposed to titanium-alloy particulate debris in vitro. *J Bone Joint Surg Am* 1999;81:603-615.

73. Shanbhag AS, Cho DR, Choy BK, et al: The transcriptional response program of human monocyte activation by polyethylene. *Trans Orthop Res Soc* 2000;25:52.

Are there biologic markers of wear?

In seeking laboratory tests that relate to implant wear, both the products of the wear process and markers that reflect the biologic consequences of wear should be considered. A few basic concepts related to the development and interpretation of laboratory tests will also be addressed.

Products of Wear

Abrasive, adhesive, and fatigue wear produces particles and ions that reflect the composition of the articulating surfaces. These include metals, ceramics, polyethylene, polymethylmethacrylate (PMMA) and its monomers, radiographic contrast media (eg, barium sulfate), corrosion products, and the consequences of implant surface treatments, such as calcium phosphates and grit blast media.

Particles of orthopaedic wear debris can be identified in tissues around implants. Although the migration of particles to regional lymph nodes and distant organs has been well documented,[1] these particles are difficult to measure without a biopsy or autopsy. Furthermore, particles of wear debris can be confused with particles of endogenous pigments, crystals, or debris from other sources; the sensitivity and specificity provided by light microscopic evaluation of remote tissues is surprisingly low.[2,3] Microanalytic tests (eg, infrared spectroscopy, energy dispersive x-ray spectroscopy, and electron microprobe analysis) can increase diagnostic specificity, but the invasive nature of tissue sampling suggests that particle detection is a relatively impractical way to measure implant wear or its consequences.

Leaching or dissolution of polymers or soluble coatings can result in rapid dissemination of the soluble materials through the circulatory system.[4] Experimental studies have suggested that insertion of freshly mixed PMMA results in peak venous monomer concentrations within several minutes, followed by exhalation of monomer from the lungs a few minutes later.[4] Elements dissolved from calcium-phosphate implant coatings also would be expected to be disseminated rapidly, but would be undetectable without experimental labeling.

Ions from metal implants can be measured in serum and in urine. Stulberg and associates[5] demonstrated high levels of titanium in serum, urine, and synovial fluid in a group of patients with failed metal-backed patellar components of total knee prostheses. Results from Jacobs and coworkers suggest that patients with loose titanium hip implants may have higher serum titanium levels than control patients,[6] and that ions can be detected in the serum of some patients who have well-functioning implants.[7] The distribution of ions in the circulation, their concentration in tissues, and patterns of urinary excretion are complex, however, and are influenced by carrier proteins, tissue uptake, and renal excretion kinetics.[4,8] At any given time, the serum or urinary level of an ion is not directly proportional to simply the magnitude or rate of metal release from the implant, but also reflects the transport kinetics of the ion at the time of sampling. The complexity of ion transport, binding, and excretion suggests that neither serum nor urinary ion levels provide a direct measure of implant wear. Further studies correlating serum and urine ion levels with patterns of implant corrosion and wear are needed to establish the utility of these tests as markers of implant degradation.

The presence of metal ions in the urine of patients with metal implants suggests that renal excretion might help limit the systemic accumulation of products of wear. Therefore, patients with chronic renal failure might be at a higher risk for developing systemic consequences of implant wear than patients with normal renal function. Patients on chronic renal hemodialysis, for example, may consume relatively large amounts of aluminum and may develop systemic aluminum toxicity, often manifested as aluminum-associated osteomalacia. A prospective study of joint replacement patients who require renal hemodialysis for kidney disease might help define the organ systems most likely to be damaged by the products of implant wear.

Consequences of Wear

In addition to direct measurement of the products of wear, systemic tests can also provide information about the consequences of wear. Orthopaedic wear debris might have both local and distant consequences. One local consequence of wear debris is an inflammatory reaction that involves macrophages and giant cells, and may result in bone resorption. Systemic consequences of wear debris are not as well defined, but are likely to be organ-specific.

The cell biology of the inflammatory reaction to particles of orthopaedic wear debris has been studied extensively. A number of different inflammatory mediators appear to be involved, including interleukin-1, interleukin-6, tumor necrosis factor alpha, and prostaglandin E2. These substances are rapidly degraded, and their distribution is usually limited to the area of the inflammation. Therefore, while assays of local tissue or synovial fluid for these substances might provide an index of the inflammatory reaction, serum or urine cytokine levels have not yet been proven to correlate with tissue injury around implants. Instead of quantifying the cytokines, another approach is to evaluate the cells that participate in the inflammatory reaction. Lee and associates[9] compared the in vitro response to particles of wear debris of peripheral blood monocytes collected from 14 patients with total hip prostheses and 10 normal volunteers. Although cells from all individuals induced cytokine release after incubation with particles, the monocytes from patients with hip implants may have been "sensitized" to the debris. There were no significant differences in cell response between patients with osteolysis and those without osteolysis.

Another consequence of particle-induced inflammation is bone resorption, which is often coupled to new bone formation. Accordingly, an investigation of markers of bone remodeling as indices of particle-induced resorption might be useful. Many laboratory tests have been studied in an attempt to identify markers useful in following patients with osteoporosis or Paget's disease. Recent research suggests that markers of bone resorption, especially serum levels of the C-terminal telopeptide fragments of Type I collagen (CTX), are more useful than markers of bone formation, such as the N-terminal midfragment of serum osteocalcin.[10-12] The considerable individual variance in these values, however, makes interpretation of these tests quite complex. By plotting test sensitivity versus specificity (receiver operating curves), cutoff levels have been calculated that are of benefit in identifying patients who fail to respond to treatment. Delmas and colleagues[11] calculated that for a given patient, a 53% decrease in serum CTX 6 months after the initiation of hormone replacement therapy had a 90% predictive value for a measurable increase in bone mineral density 1.5 years later. These predictive values have been validated only for patients with osteoporosis being treated with hormone replacement therapy.

Markers of bone remodeling also have been used to follow patients with clinically stable Paget's disease. Alvarez and associates[13] examined the biologic variation of serum bone alkaline phosphatase, procollagen type 1 N-terminal propeptide, beta-telopeptide carboxyterminal of type I collagen, and telopeptides of carboxy- and amino terminals of type I collagen in urine. There was considerable variability among patients and a diurnal variation was identified, reflecting the circadian rhythm of bone resorption. Serum markers showed less variability than did urinary markers, and markers of bone resorption showed less within-subject variability than did markers of bone formation. Among resorption markers, the serum CTX appeared to be the most useful single marker. The overall variability of all markers, however, suggested that for Paget's disease, absolute values in serum or urine have relatively low diagnostic value, but may be useful to follow individual patients over time.[13]

One recent study attempted to use one of these markers—urinary aminoterminal telopeptides of type I collagen (NTX)—to diagnose and assess treatment of periprosthetic osteolysis.[14] Urinary NTX levels were compared in 10 patients with periprosthetic hip osteolysis, 8 patients with hip implants but without osteolysis, and 7 age-matched controls without implants. Ten additional patients with periprosthetic osteolysis were evaluated before and after alendronate treatment. Although there was no correlation between the urinary NTX level and the radiographic extent of osteolysis, patients with osteolysis had significantly higher levels of NTX than either the unimplanted control group or the group of patients with hip implants but no osteolysis. In addition, there was a significant decrease in urinary NTX in 9 of 10 patients who received a 6-week course of alendronate treatment for periprosthetic osteolysis.[14] Despite the limitations of these studies, it appears that measuring indices of bone resorption, especially serum and urinary NTX or CTX, may provide markers of at least one consequence of implant wear.

Systemic consequences of implant wear are not as well defined as localized osteolysis, but could be either organ-specific or generalized. Organ-specific toxicity could be identified by clinical findings in conjunction with appropriately targeted laboratory tests. Several recent studies[15,16] using cytogenetics have suggested that patients with implants may experience more generalized cellular toxicity, a topic that needs further investigation.

Basic Laboratory Test Concepts

Optimal use of expensive laboratory tests requires an understanding of the limitations of each test as well as the prevalence of a disorder in the population under study. The terms "sensitivity" and "specificity" are used to help define the most appropriate use of a test. Sensitivity is the prevalence of true-positive results when a test is applied to patients who are known to have the disease; it is the number of true-positive results divided by the sum of the true-positive and false-negative results.[17] Specificity describes the prevalence of true-negative results when a test is applied to patients who are at risk for the disease but are known to be free of it; it is defined as the number of true-negative results divided by the sum of the true-negative and false-positive results. In general, sensitivity is increased at the expense of specificity. Tests of high sensitivity are used for screening, and may be followed by tests of higher specificity. Sensitivity and specificity must be determined in the context of the appropriate patient population. If screening tests are applied to a healthy population, then false-positive results will occur. For example, if joint fluid is aspirated from a series of consecutive patients who have a total joint implant, regardless of physical findings, the high rate of false-positive bacterial culture results suggest that aspiration of joint fluid is a poor test for infection.[18] On the other hand, if aspiration and culture of joint fluid is performed only for patients who have a prosthesis and physical findings suggestive of infection, the test is much more reliable. The predictive value of a positive test result (ie, the percentage of all positive tests that are true-positives) also varies with the prevalence of the disease. Laboratory tests have a high predictive value when a disorder is very common, and a low predictive value when a disorder is rare.[17] These concepts should be taken into consideration when selecting control populations for prospective studies related to implant wear. At this time, insufficient published data are available to calculate the sensitivity, specificity, or predictive value of any laboratory tests with reference to implant wear or bone resorption.

Relevance

Although the extent of implant wear can be inferred from radiographic studies, these studies have relatively low sensitivity and may not correlate with the biologic consequences of wear in any given patient. If laboratory tests were developed that could provide indices of either implant wear itself, or of the biologic consequences of wear (eg, bone resorption), such tests might be used to identify patients at risk for implant loosening or bone resorption, provide indices reflecting the rate of progression of bone resorption, and assess efficacy of pharmacologic or other treatments.

Future Directions for Research

Additional studies are needed to further document systemic distribution of ions in patients with total joint replacements of different types (as well as in suitable control patients) in correlation with patterns of implant wear and degradation; document the extent, distribution, and consequences of systemic ions in patients with chronic renal failure who have total joint replacements; and test the predictive value of cytokine levels obtained from serum and synovial fluid with respect to implant wear.

Research should also be conducted to determine reference ranges for markers of bone turnover, especially serum and urine N-terminal and C-terminal telopeptides of Type I collagen (CTX and NTX), in patients with clinically satisfactory implants, patients with loose implants, patients with osteolysis but stable implants, and control patients (eg, those with osteoarthritis but no joint replacements).

References

1. Urban RM, Jacobs JJ, Tomlinson MJ, Gavrilovic J, Black J, Peoch M: Dissemination of wear particles to the liver, spleen and abdominal lymph nodes of patients with hip or knee replacement. *J Bone Joint Surg Am* 2000;82:457-475.
2. Bauer TW: Editorial: Identification of orthopaedic wear debris. *J Bone Joint Surg Am* 1996;78:479-481.
3. Shea KG, Bloebaum RD, Avent JM, Birk GT, Samuelson KM: Analysis of lymph nodes for polyethylene particles in patients who have had a primary joint replacement. *J Bone Joint Surg Am* 1996;78:497-504.
4. Black J: Biological performance of materials, in *Fundamentals of Biocompatibility*, ed 3. New York, NY, Marcel Dekker, 1999, pp 259-311.
5. Stulberg BN, Merritt K, Bauer TW: Metallic wear debris in metal-backed patellar failure. *J Applied Biomat* 1994;5:9-16.
6. Jacobs JJ, Skipor AK, Black J, Urban RM, Galante JO: Release and excretion of metal in patients who have a total hip replacement component made of titanium-base alloy. *J Bone Joint Surg Am* 1991;73:1475-1486.
7. Jacobs JJ, Skipor AK, Patterson LM, et al: Metal release in patients who have had a primary total hip arthroplasty: A prospective, controlled, longitudinal study. *J Bone Joint Surg Am* 1998;80:1447-1458.
8. Woodman JL, Jacobs JJ, Galante JO, Urban RM: Metal ion release from titanium-based prosthetic segmental replacements of long bones in baboons: A long-term study. *J Orthop Res* 1984;1:421-430.
9. Lee S-H, Brennan FR, Jacobs JJ, Urban RM, Ragasa DR, Glant TT: Human monocyte/macrophage response to cobalt-chromium corrosion products and titanium particles in patients with total joint replacements. *J Orthop Res* 1997;15:40-49.
10. Bjarnason NH, Christiansen C: Early response in biochemical markers predicts long-term response in bone mass during hormone replacement therapy in early postmenopausal women. *Bone* 2000;26:561-569.

11. Delmas PD, Hardy P, Garnero P, Dain MP: Monitoring individual response to hormone replacement therapy with bone markers. *Bone* 2000;26:553-559.

12. Riggs BL: Are biochemical markers for bone turnover clinically useful for monitoring therapy in individual osteoporotic patients? *Bone* 2000;26:551-552.

13. Alvarez L, Ricos C, Peris P, et al: Components of biological variation of biochemical markers of bone turnover in Paget's bone disease. *Bone* 2000;26:571-576.

14. Antoniou J, Huk O, Zukor D, Eyre D, Alini M: Collagen crosslinked N-telopeptides as markers for evaluating particulate osteolysis: A preliminary study. *J Orthop Res* 2000;18:64-67.

15. Stea S, Visentin M, Granchi D, et al: Sister chromatid exchange in patients with joint prostheses. *J Arthroplasty* 2000;15:772-777.

16. Case CP, Langkamer VG, Howell RT, et al: Preliminary observations on possible premalignant changes in bone marrow adjacent to worn total hip arthroplasty implants. *Clin Orthop* 1996;329:S264-S279.

17. Galen RS, Gambino SR: Beyond normality, in *The Predictive Value and Efficiency of Medical Diagnoses*. New York, NY, John Wiley & Sons, 1975.

18. Barrack RL, Harris WH: The value of aspiration of the hip joint before revision total hip arthroplasty. *J Bone Joint Surg Am* 1993;75:66-76.

What potential biologic treatments are there for osteolysis?

The development of osteolysis is influenced by a number of factors, which can operate alone or in tandem. The cellular response to wear particles, known to be a potent cause of osteolysis, has been investigated. The mechanical effect of the prosthesis can have a marked effect on the bone cell response, producing profound bone loss within a few weeks of implantation. Contamination with endotoxin and overt infection are also potent causes of periprosthetic osteolysis.

One immediate cause of bone loss, often ignored in clinical studies, is effected by the surgeon who prepares the bone at the time of initial implantation. Prosthesis stability also has a very strong influence on the development of osteolysis. Instability affects the accumulation of wear products at the interface and the strain rates imposed on the adjacent cells. These local effects need to be considered in the context of systemic bone dynamics—the rate of bone loss in the whole skeleton, which is particularly relevant in osteoporotic, elderly women and patients who have disuse atrophy from enforced rest or inactivity.

Osteolysis takes different forms. It can be very localized, eg, the lacunae that contain wear particles and phagocytic cells are found close to the surface of the implant. A more widespread reduction in bone mass affecting the proximal femur is frequently seen after implantation of an otherwise successful hip prosthesis, due in part to differences in material properties (such as elasticity) of the bone and prosthesis. The general level of osteoporosis may have a profound influence on this phenomenon.[1,2]

The effect of wear particles has been extensively studied. Movement of a metal prosthesis against bone can cause osteolysis.[3] In the presence of particles, when the movement was stopped the osteolytic lesion did not heal; if there were no particles in the osteolytic lesion, the bone defect healed. This clearly indicates the symbiosis between different mechanisms causing bone loss. Kobayashi and associates[4] studied retrieved specimens from revision arthroplasties and demonstrated that a specific concentration of particles is required in order to produce or maintain an osteolytic lesion adjacent to a prosthesis.

With respect to bone loss at the bone-prosthesis interface, the mechanical effect of an unstable prosthesis ranges from mild to disastrous. Furthermore, implantation of a prosthesis has a considerable effect on bone activity.[5-9] Although radioisotope techniques provide good indication of bone activity, the metabolic changes caused by mechanical effects must be differentiated from those caused by infection.[10]

The use of dual-energy x-ray absorptriometry to detect bone density related to Gruen zones has yielded a remarkable insight into the effect of a successful implantation in the proximal femur.[11] Further, the continuing loss of bone postoperatively was not uniformly distributed; by 12 months, the overall pattern of change in bone mineral density was that of resorption that increased from the distal to the proximal part of the femur. Increase in bone density was seen below the tip of the prosthesis.[12] The distribution of change in bone mineral density in the proximal femur may have a number of explanations. It may follow the postoperative change in the strain pattern being transmitted to the bone, or it may represent the effect of destruction of the complex trabecular pattern within the proximal femur.[13-15]

Increasing exercise plays an important role in bone mineral formation. Cohen and associates[16] demonstrated by skeletal measurements that a group of 17 male college oarsmen produced a significant 2.9% increase in bone mineral density in the lumbar spine and a 4.2% increase in bone mineral content in the same region after 7 months of exercise. Other authors have confirmed that strain affects bone formation.[17] The effect of strain on bone cell metabolism can also be demonstrated in tissue culture.[18,19] Changing the strain on the cells, or the frequency by which the strain is applied, can lead to either an increase in bone formation and mineralization or in the removal of bone (such as when an arterial aneurysm is in contact with bone, a vertebra, or a rib).

Although the degree of osteoporosis of the bone at the time of implantation is very important for prosthetic survival, possibly even more important is the continued development of osteoporosis throughout the working lifetime of the artificial joint. Endosteal bone removal and periosteal bone deposition is characteristic of the osteoporotic process. This has the effect of making the bony tube wider with thinner walls. Unless there is some effective bone remodeling at the bone-prosthesis interface, the prosthesis will loosen, leading to failure. In many total joint replacement patients, osteoporosis is probably not reversible but may be arrested by medications.

Surgeons are very aware of the devastating results of infection and take all appropriate precautions to prevent contamination at the time of surgery. Incorporating antibiotics in the bone cement can reduce the consequences of infection. Endotoxin contamination may be underestimated as a cause of prosthetic failure. This contamination can be introduced at any stage of manufacture or packaging.

Potential Treatment Methods

The reduction of the number of wear particles produced at the interface is likely to produce a major improvement in the survival of total joint replacements. Preventing the ingress of particles around the interface may reduce the incidence of localized osteolysis due to wear debris. The use of bone cement effectively seals the interface at operation, in contrast to the implantation of a cementless prosthesis with which no satisfactory particle-resistant seal is initially available. This is particularly important because an increased incidence of particle generation is seen immediately after implantation of the prosthesis, due to the polishing process as the articulation is "run in". The interface may be sealed using other methods, for example hydroxyapatite coatings, tissue glue, or occlusive films made from synthetic material or from the patient's own tissue.[20]

The use of specific drugs may reduce bone loss at the implant site. Systemic bisphosphonates have been shown to preserve fracture toughness of bone in rabbit experiments.[21] Astrand and Aspenberg[22] demonstrated that alendronate did not inhibit instability-induced bone resorption. In this study using rats, a controllable but unstable implant was used as a model to compare the effect of alendronate delivered via minipumps. The alendronate increased the ash weight of the proximal metaphyses of the tibia by more than 40%, but there was no inhibition of the instability-induced bone resorption.

Shanbhag and associates[23] demonstrated in a canine model that bisphosphanate therapy for 24 weeks effectively inhibited osteolysis around hip prostheses that had been exposed to wear particles but which were otherwise stable. Bisphosphonate drugs and their derivatives, together with other compounds that reduce the function of osteoclasts, may be useful in the reduction of bone loss secondary to osteolysis. However, if biophosphonates are used around the time of surgery, the early remodeling phase of bone may be prevented. Intermittent use of this type of pharmaceutical agent may be indicated in some patients. Some biophosphonates have been proven to be effective in reducing the rate of osteoporotic fractures after an initial fracture of the neck of the femur; however, these patients do not commonly suffer from degenerative arthritis of the hip.

Allograft cancellous bone chips may be used to replenish areas of bone loss in patients with stable cementless acetabular components.[24]

Relevance

The response of the supporting tissue to the insertion of prostheses is of prime importance to the long-term performance of joint replacements. Modifications to this environment may prove very advantageous. Because of the complexity of interrelated factors, however, it is unlikely that a single intervention will produce a miraculous cure.

Future Directions for Research

Careful study of the periprosthetic environment should lead to better understanding of the detailed changes at the interface. A combination of different investigation methods is likely to be most productive, especially if the images can be coregistered.

Prosthetic design should incorporate the need to match prosthetic modulus with bone and mechanical strategies to prevent wear particles from reaching the interface.

The surveillance of implanted joints should include long-term multicenter survival analysis. Detailed study of smaller numbers of patients using multiple investigation methods may be very effective in detecting early failure of new joint replacements.

References

1. Kroger H, Miettinen H, Arnala I, Koski E, Rushton N, Suomalainen O: Evaluation of periprosthetic bone using dual x-ray absorptiometry: Precision of the method and effect of operation on bone mineral density. *J Bone Miner Res* 1996;11:1526-1530.
2. Kroger H, Vanninen E, Overmyer M, Miettinen H, Rushton N, Suomalainen O: Periprosthetic bone loss and regional bone turnover in uncemented total hip arthroplasty: A prospective study using high resolution single photon emission tomography and dual energy x-ray absorptiometry. *J Bone Miner Res* 1997;12:487-492.
3. Aspenberg P, Herbertsson P: Periprosthetic bone resorption: Particles versus movement. *J Bone Joint Surg Br* 1996;78:641-646.
4. Kobayashi A, Freeman MA, Bonfield W, et al: Number of polyethylene particles and osteolysis in total joint replacements: A quantitative study using a tissue-digestion method. *J Bone Joint Surg Br* 1997;79:844-848.
5. Cohen B, Rushton N: A comparative study of peri-prosthetic bone mineral measurement using two different dual-energy x-ray absorptiometry systems. *Br J Radiol* 1994;67:852-855.
6. Cohen B, Rushton NL: Accuracy of DEXA measurement of bone mineral density after total hip arthroplasty. *J Bone Joint Surg Br* 1995;77:479-448.
7. Rushton N, Wraight EP: Technetium 99m methylene diphosphonate scanning in Thompson hemiarthroplasties. *Br J Radiol* 1980;53:781-783.
8. Rushton N, Wraight EP: The early detection of acetabular migration after Thompson hemiarthroplasty using technetium bone scanning. *Nucl Med Commun* 1981;2:345-350.
9. Weiss PE, Mall JC, Hoffer PB, Murray WR, Rodgrigo JJ, Genamat HK: 99MTC-methylene bisphosphanate bone imaging in the evaluation of total hip prostheses. *Radiology* 1979;133:727-730.
10. Rushton N, Coakley AJ, Tudor J, Wraight EP: The value of technetium and gallium scanning in assessing pain after total hip replacement. *J Bone Joint Surg Br* 1982;64:313-331.
11. Gruen TA, McNice GM, Amstutz HC: Modes of failure of cemented stem type femoral components: A radiographic analysis of loosening. *Clin Orthop* 1979;141:17-27.
12. Cohen B, Rushton N: Bone remodeling in the proximal femur after Charnley Total Hip Arthroplasty. *J Bone Joint Surg Br* 1995;77:815-819.
13. Field RE, Dixon AK, Lawrence JP, Rushton N: Bone density distribution within the femoral head and neck: An examination by high resolution computed tomography. *Skeletal Radiol* 1990;19:319-325.
14. Ruben CT: Skeletal strain and the functional significance of bone architecture. *Calcif Tissue Int* 1979;36:S11-S18.
15. Rubin CT, Lanyon LE: Regulation of bone formation by applied dynamic loads. *J Bone Joint Surg Am* 1984;66:397-402.
16. Cohen B, Millett PJ, Mist B, Laskey MA, Rushton N: Effect of exercise training programme on bone mineral density in novice college rowers. *Br J Sports Med* 1995;29:85-88.

17. Goodship AE, Lanyon LE, McFie H: Functional adaption of bone to increased stress: An experimental study. *J Bone Joint Surg Am* 1979;61:539-546.

18. Murray DW, Rushton N: The effect of strain on bone cell prostaglandin E2 release: A new experimental method. *Calcif Tissue Int* 1990;47:35-39.

19. Yeh C, Rodan A: Tencile forces enhanced PGE synthysis in osteoblasts grown on collegen ribbon. *Calcif Tissue Int* 1989;36:S67-S71.

20. Bhumbra RP, Walker PS, Berman AB, Emmanual J, Barrett DS, Blunn GW: Prevention of loosening in total hip replacements using guided bone regeneration. *Clin Orthop* 2000;372:192-204.

21. Bellingham CM, Lee JM, Moran EL, Bogoch ER: Bisphosphonate (Pamidronate/APD) prevents arthritis-induced loss of fracture toughness in the rabbit femoral diaphysis. *J Orthop Res* 1995;13:876-880.

22. Astrand J, Aspenberg P: Alendronate did not inhibit instability-induced bone resorption: A study in rats. *Acta Orthop Scand* 1999;70:67-70.

23. Shanbhag AS, Hasselman CT, Rubash HE: Inhibition of wear debris mediated osteolysis in a canine total hip arthroplasty model. *Clin Orthop* 1997;344:33-43.

24. Maloney WJ, Herzwurm P, Paprosky W, Rubash HE, Engh CA: Treatment of pelvic osteolysis associated with a stable acetabular component inserted without cement as part of a total hip replacement. *J Bone Joint Surg Am* 1997;79:1628-1634.

Material and Design Considerations

What design and material factors influence wear in total joint replacement?

What is the role of wear testing and joint simulator studies in discriminating among materials and designs?

What design factors influence wear behavior in total knee replacement?

What design factors influence wear behavior at the articulating surfaces in total hip replacement?

What are the wear mechanisms and what controls them?

What material properties and manufacturing procedures influence wear mechanisms?

What modifications can be made to materials to improve wear behavior?

What evidence is there for using alternative bearing materials?

What design and material factors influence wear in total joint replacement?

Wear in total joint replacements is caused by relative motion under load at articulating surfaces or at interfaces between modular components. Wear is the removal of material from the surface, fundamentally a mechanical process; the stresses associated with surface damage exceed the strength of the material and particles are liberated.

Wear that occurs at articulating surfaces of metal-on-polyethylene joint replacements has received the most attention in recent years. Wear at metal-polyethylene interfaces of modular components is probably avoidable with designs that eliminate or at least minimize relative interface motion. In metal-on-polyethylene articulating surfaces of hip and knee replacements, however, relative motion is unavoidable and loads are large.

Surface damage is a mode of structural failure of the bone-implant system. The stresses associated with damage are a complex interaction of implant design variables, patient characteristics, and surgical factors (Fig. 1). Implant design variables are controlled variables; they include choice of component geometry, material, manufacturing processes, and sterilization methods. Patient factors, such as weight, activity, and bone properties, are uncontrolled environmental variables. Surgical factors are a combination of design and environmental variables—the desired position and orientation of the implanted components are design variables, while the normal variations

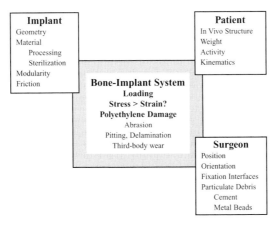

Figure 1 *Many factors affect wear in vivo, and influence one another as well.*

about these desired positions are environmental variables. In metal-on-polyethylene bearings, most of the damage occurs in the polyethylene. The surface strength of polyethylene is affected by complex interactions of design and environmental factors, eg, resin type, sterilization method, shelf aging, loading, and joint kinematics.

Determination of the influence of specific factors on wear in total joint replacements requires an understanding of the mechanisms of damage. These mechanisms are still not well known for polyethylene, but progress is being made. Circumstantial evidence is used primarily to relate calculated stresses to damage observed in vivo or in joint simulation. Furthermore, because surface damage is a function of the characteristics of a complex bone-implant system, it is extremely difficult to isolate the effects of a single design or material factor on wear. The impact of design and material factors on wear in total joint replacements is complicated. Influences of the environmental variables on structural behavior may be greater than the impact of design changes. Improved performance will require design and material changes that exert greater influence than the environmental variables on wear.

Wear simulators can eliminate environmental variables associated with surgical factors and some patient factors. Simulators are not simply material screening devices; rather, they are tribologic testing systems. The performance of the implant specimen in the simulator is a complex interaction of many variables (Fig. 2).

Although design and material factors associated with wear are always interacting, general factors can be identified that are associated either with the magnitudes of the stresses produced by joint contact or with the strength of the polyethylene at or near the surface. The factors that have been shown to influence the stress magnitudes are component thickness, polyethylene stress-strain behavior, bearing conformity, and cyclic loading. Material factors associated with strength are more difficult to identify because the mechanisms of microscopic or macroscopic damage remain largely unknown.

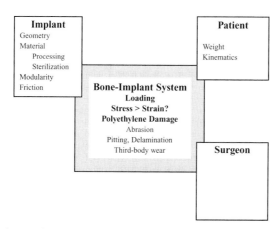

Figure 2 Simulators eliminate surgical variables and some patient variables.

Association of stress measures with damage has been primarily circumstantial. Recently, however, large deformation behavior—in particular work to failure, or toughness—of polyethylene has been correlated with hip simulator wear for acetabular cups and with damage scores of retrieved tibial components for total knee replacements.[1,2]

Factors Influencing Stresses

A general view of the problem of stress and component wear can be provided in an axisymmetric model of normal, frictionless contact between rigid metallic and polyethylene components (Fig. 3). If the radii of the spherical contact surfaces are properly selected, the stresses and strains from the axisymmetric model closely approximate those of three-dimensional models of contemporary implants (Fig. 4).[3,4] Therefore, the general problem can be considered using the simpler axisymmetric model, which shows that stresses and strains increase with decreasing thickness, decreasing conformity, and increasing modulus of the polyethylene (Figs. 5 and 6). These figures show contact stresses, but other stress measures may be more directly associated with fatigue failure, including the range of maximum principal stress, maximum shear stress, and maximum von Mises stress. These stress measures generally increase when the contact stresses increase.

In many polyethylene knee components, the predominant failure mode is fatigue due to cyclic loading of the components, particularly in degraded implants. In hip components, however, the predominant failure mode is abrasive-adhesive wear because they are more conforming. Similar axisymmetric analyses for acetabular cups show that stresses are smaller than those in knees and deformations are in the elastic range. In the limit, if perfect conformity is achieved, stresses become independent of thickness and elastic modulus.[5] This result is consistent with the observation that after a peri-

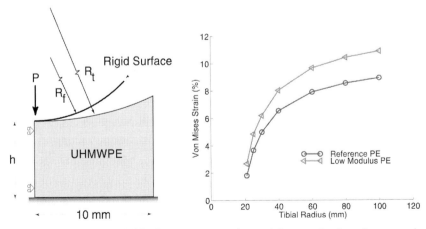

Figure 3 *Axisymmetric model for knee contact, and material properties for reference and low-modulus polyethylene.*

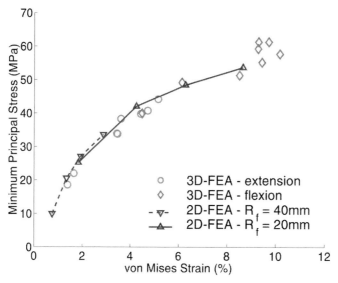

Figure 4 *Axisymmetric model of knee joint contact. Results from axisymmetric models approximate those of three-dimensional models. (Reproduced with permission from Bartel DL, Rawlinson JJ, Burstein AH, Ranawat CS, Flynn WF Jr: Stresses in polyethylene components of contemporary total knee replacements. Clin Orthop 1995;317:76-82.)*

od of wear-in or creep, the in vivo wear rates for acetabular cups are constant.

Cyclic loading also affects the stresses associated with fatigue failure of the polyethylene. Stresses at a point in the polyethylene near or at the surface vary during a particular activity, such as gait, and from activity to activity. A damage function can be defined to account for the amplitudes and frequencies of the maximum shear stress cycles that occur at a point on the surface and then used to predict fatigue failure.[6] When the loading and unloading behavior of the polyethylene is included in models of polyethylene com-

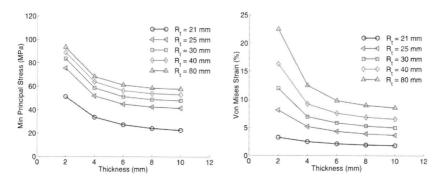

Figure 5 *Stresses and strains increase when thickness or conformity decreases. R is the radius of the polyethylene surface; as R increases, the surfaces between the metal and polyethylene become less conforming and stresses increase.*

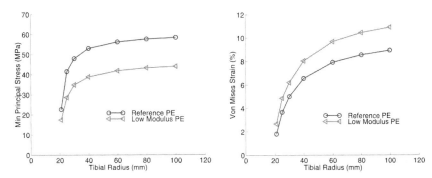

Figure 6 *Stresses and strains increase when conformity is decreased (tibial radius increases) or the modulus of elasticity is increased.*

ponents, residual stresses are created that are tensile at the surface and add to the tensile stresses due to the recurring joint load.[7] Damage accumulation in a nonlinear stress field can also be studied by modeling crack propagation.[8] Cyclic loading therefore affects stresses in two ways. First, the stress magnitudes associated with damage, such as the maximum principal stress, are increased because of the residual stresses; this could be considered a type of damage accumulation in itself. Second, damage can accumulate as cracks propagate.

Material Factors

If the stresses that are associated with damage could be estimated, a strength criterion would be useful for design purposes, especially if such a criterion could be combined with a simple test to determine if one polyethylene were more damage resistant than another. Although failure theories based on maximum shear stress or maximum distortion energy have been used for design of other engineered structures and machine components, they have not been useful for polyethylene component design, possibly because surface damage in polyethylene components involves large deformations and multiaxial loading of the material. Particle generation in acetabular cups is attributed to processes in which strands of polyethylene are drawn from the surface and then sheared off. Fatigue crack propagation in tibial components for total knee replacements is affected by the large deformations accompanying nonconforming contact that lead to residual stresses.[8-10] As a result, the simpler correlations between failure or damage and the results of linear uniaxial testing, or even tests based on fracture mechanics, have been inadequate for associating material characteristics with stress states. Because a strength factor has not yet been identified, material modifications have generally been tested and screened based on retrieval analysis and joint simulators. Such studies have defined the important roles of quality control in manufacturing, shelf aging, and crosslinking in wear reduction, but have not provided a strength criterion that can be used for design.

One promising material factor is work to failure, or toughness, as determined by the small punch test.[11] This test (Fig. 7) can be used to determine work to failure, peak load, ultimate load, and ultimate displacement (Fig. 8).[12] Using finite element stress analysis in conjunction with the test, peak strength, ultimate strength, and ultimate strain can also be determined. This test is attractive because the specimens are small enough that properties can be determined as a function of depth through actual joint components, enabling precise measures of heterogeneous changes such as degradation to be made. The test also involves large deformations and multiaxial loading

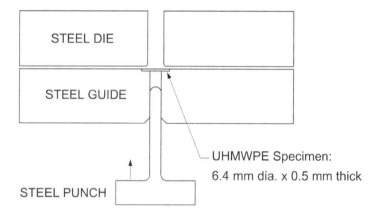

Figure 7 *Schematic of the small punch test apparatus. (Reproduced with permission from Kurtz SM, Foulds JR, Jewett CW, Srivastav S, Edidin AA: Validation of a small punch testing technique to characterize the mechanical behaviour of ultra-high-molecular-weight polyethylene. Biomaterials 1997;18:1659-1663.)*

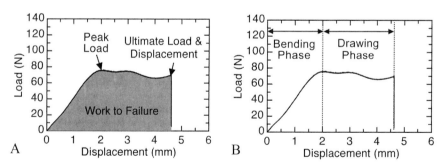

Figure 8 A, *Load-displacement curve for polyethylene from the small punch test is characterized by an initial peak load, ultimate load, ultimate displacement, and work to failure. Before the initial peak load, deformation of the specimen occurs primarily in bending. **B,** After the initial peak load, deformation of the specimen is dominated by drawing or stretching of the specimen around the head of the punch in a state of equibiaxial tension. (Reproduced with permission from Edidin AA, Kurtz SM: Development and validation of the small punch test for UHMWPE used in total joint replacements, in Katsube N, Soboyejo WO, Sacks N (eds): Functional Biomaterials. Zurich, Switzerland, Trans Tech Publications, 2001.)*

142

consistent with the deformations and loading associated with observed failure modes in total joint replacements.

The work to failure from the small punch test has been correlated with hip simulator wear rates and with the amount of observed damage in retrieved tibial components. The inverse relationship between hip simulator wear and work to failure for uncross-linked polymers demonstrates that the least work to failure is associated with the greatest wear rates (Fig. 9).[1,13] However, for increased cross-linking by repeated doses of gamma radiation in nitrogen (but without further thermal processing), the material with the least work to failure was associated with the lowest wear rates. A similar correlation between work to failure and damage has been observed for retrieved tibial components (Fig. 10).[2,14] Thus, work to failure correlates with both abrasive-adhesive wear in the conforming geometries of hip joints and the fatigue damage mechanisms observed in less conforming total knee replacements.

Relevance

Given the current mature state of total joint implant design, the greatest reduction in wear will probably be facilitated by increasing damage resistance of the material or maintaining the strength of the virgin material over

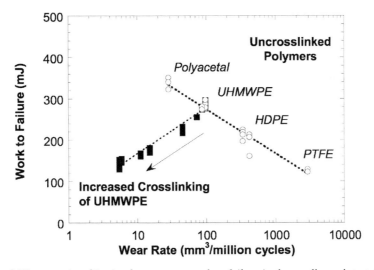

Figure 9 *Wear rate in a hip simulator versus work to failure in the small punch test. For uncross-linked polymers (open circles), a decrease in wear rate was associated with increased work to failure (closed squares). The opposite trend was observed when polyethylene was cross-linked by repeated doses of gamma radiation in nitrogen, but without subsequent annealing (black squares). (Adapted with permission from Edidin AA, Kurtz SM: Influence of mechanical behavior on the wear of 4 clinically relevant polymeric biomaterials in a hip simulator. J Arthroplasty 2000;15:321-331 and Edidin AA, Pruitt L, Jewett CW, Crane DJ, Roberts D, Kurtz SM: Plasticity-induced damage layer is a precursor to wear in radiation-cross-linked UHMWPE acetabular components for total hip replacement: Ultra-high-molecular-weight polyethylene. J Arthroplasty 1999;14:616-627.)*

time, rather than design changes to reduce stresses due to contact. The ranges over which stresses are relatively insensitive to thickness are somewhat less restrictive than those from previous studies because more recent work better incorporates the nonlinear characteristics of polyethylene. Conformity should be maximized to reduce stresses, but there are limits to the conformity that can be tolerated in fixed-bearing knee joints. If knee replacements are too conforming, there will be insufficient laxity to allow soft tissue to share in the resistance to axial torsion, putting the fixation interfaces at risk. A major goal of mobile-bearing knees is to allow more conforming femoral-tibial contact without reducing torsional laxity. The disadvantage of mobile bearings is that additional interfaces are introduced at which wear can occur; although surface fatigue may be reduced, abrasive-adhesive wear may increase. Finally, the different stresses that result from surgical or inter- and intrapatient variations in environmental variables may be substantial. The relative influence of design and environmental variables should be considered when designs are modified.

Future Directions for Research

Even when designs of the articulating surfaces are optimized, they can be undermined by lack of attention to the design of metal backings, polyethylene liner capture and locking mechanisms, and surgical features that increase stresses, such as acetabular cup orientation.[15] Design may also play

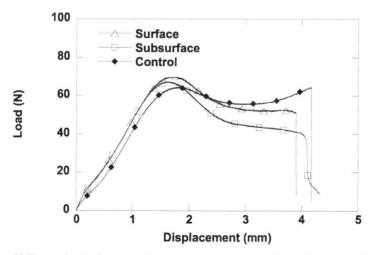

Figure 10 The work to failure of a tibial component gamma-radiated in air and shelf-aged for 5 years showed reduced work to failure in the subsurface, consistent with in vivo damage observed in retrieval analyses. The subsurface region is also the location of maximum shear stress and von Mises stress and strain. (Reproduced with permission from Edidin AA, Jewett CW, Kalinowski A, Kwarteng K, Kurtz SM: Degradation of mechanical behavior in UHMW-PE after natural and accelerated aging. Biomaterials 2000;21:1451-1460.)

a role in other aspects of the system; components, for example, may be designed to reduce the variations from surgical environmental factors.

If a relevant material strength criterion can be identified, then the attention of design can be turned from minimizing stresses to maximizing joint function subject to strength as a constraint. Improved performance might be found at the expense of higher stresses produced in a design using thinner, less conforming components if the material is indeed stronger. Perhaps the most obvious example is the recent proposal to use thinner acetabular liners to allow larger femoral head sizes, reducing the risk of dislocation. Similar advantages could be proposed in knee replacements, for which functional improvement might require the use of less conforming designs. At the present time, however, it seems prudent to attack the problem from both ends by developing joint replacement bearings that both minimize stresses and maximize material resistance to damage.

Work to failure as determined from a small punch test is a promising metric for quantifying degradation and comparing the strengths of competing material processes and sterilization methods for polyethylene components. It has been estimated that two to four million joint replacements gamma irradiated in air are currently in varying periods of shelf aging.[16] Work to failure provides a measure for quantifying degradation observed in retrieval analyses, and should make it possible to assess this large population.

Wear of uncross-linked polymers was correlated with work to failure as one might intuitively expect; materials with large work to failure wear less (Fig. 9). When the materials were crosslinked, however, the effect was reversed. Although a joint simulator (Fig. 2) is less complex than the in vivo system (Fig. 1), nevertheless it is still a system, a tribologic testing machine, in which many factors interact. Cross-linking under certain situations could reduce work to failure, yet change the mechanism by which particles are removed from the surface because of changes to the surface chemistry that are effective in vitro but not in vivo. Further thermal processing could be used on cross-linked specimens to achieve low wear without loss of toughness.

Finally, work to failure, expressed in terms of stress and strain, may also provide a criterion that can be used for comparing designs computationally. This would reduce if not eliminate the current dependence on circumstantial evidence for coupling stress computations with damage observed in simulator specimens and retrieved components and could provide a basis for predicting damage risk in new designs.

References

1. Edidin AA, Kurtz SM: Influence of mechanical behavior on the wear of 4 clinically relevant polymeric biomaterials in a hip simulator. *J Arthroplasty* 2000;15:321-331.
2. Edidin AA, Jewett CW, Kalinowski A, Kwarteng K, Kurtz SM: Degradation of mechanical behavior in UHMWPE after natural and accelerated aging. *Biomaterials* 2000;21:1451-1460.

3. Rawlinson JJ, Armitage BM, Bartel DL: Contact stresses in UHMWPE: The need for nonlinear material models. *Trans Orthop Res Soc* 1999;24:846.

4. Bartel DL, Rawlinson JJ, Burstein AH, Ranawat CS, Flynn WF Jr: Stresses in polyethylene components of contemporary total knee replacements. *Clin Orthop* 1995;317:76-82.

5. Bartel DL, Burstein AH, Toda MD, Edwards DL: The effect of conformity and plastic thickness on contact stresses in metal-backed plastic implants. *J Biomech Eng* 1985;107:193-199.

6. Sathasivam S, Walker PS: Computer model to predict subsurface damage in tibial inserts of total knees. *J Orthop Res* 1998;16:564-571.

7. Estupinan JA, Bartel DL, Wright TM: Residual stresses in ultra-high molecular weight polyethylene loaded cyclically by a rigid moving indenter in nonconforming geometries. *J Orthop Res* 1998;16:80-88.

8. Estupinan JA, Bartel DL, Wright TM: Simulation of surface crack propagation in UHMWPE. *Trans Orthop Res Soc* 1998;23:703.

9. Estupinan JA, Bartel DL: Residual stress field around subsurface void in UHMWPE loaded by moving indenter as site for fatigue crack propagation. *Trans Orthop Res Soc* 1996;21:496.

10. Estupinan JA, Bartel DL: Surface residual tensile stress after cyclic loading of UHMWPE by a rigid indenter. *Trans Orthop Res Soc* 1996;21:48.

11. Kurtz SM, Jewett CW, Foulds JR, Edidin AA: A miniature specimen mechanical testing technique scaled to articulating surface of polyethylene components for total joint arthroplasty. *J Biomed Mater Res* 1999;48:75-81.

12. Kurtz SM, Foulds JR, Jewett CW, Srivastav S, Edidin AA: Validation of a small punch testing technique to characterize the mechanical behaviour of ultra-high-molecular-weight polyethylene. *Biomaterials* 1997;18:1659-1663.

13. Edidin AA, Pruitt L, Jewett CW, Crane DJ, Roberts D, Kurtz SM: Plasticity-induced damage layer is a precursor to wear in radiation-cross-linked UHMWPE acetabular components for total hip replacement: Ultra-high-molecular-weight polyethylene. *J Arthroplasty* 1999;14:616-627.

14. Kurtz SM, Rimnac CM, Pruitt L, Jewett CW, Goldberg V, Edidin AA: The relationship between clinical performance and large deformation mechanical behavior of retrieved UHMWPE tibial inserts. *Biomaterials* 2000;21:283-291.

15. Kurtz SM, Edidin AA, Bartel DL: The role of backside polishing, cup angle, and polyethylene thickness on the contact stresses in metal-backed acetabular components. *J Biomech* 1997;30:639-642.

16. Edidin AA, Kurtz SM: Development and validation of the small punch test for UHMWPE used in total joint replacements, in Katsube N, Soboyejo WO, Sacks N (eds): *Functional Biomaterials*. Zurich, Switzerland, Trans Tech Publications, 2001.

What is the role of wear testing and joint simulator studies in discriminating among materials and designs?

Material selection and component design are important factors in the performance and durability of total joint replacements. Both laboratory bench tests and joint simulators have been used to evaluate the influence of materials and designs on wear. The role of wear testing and joint simulation studies is often unclear, however, in terms of discriminating the effect of materials and design on performance.

Controlled bench wear testing should be used to develop an understanding of wear mechanisms and the influence of environmental, design, and material parameters on wear behavior. Replicating the specific conditions occurring in a hip or knee joint, simulator testing can then be used to test specific design and material combinations. Used properly, bench wear testing can also serve as a screening tool to identify grossly unsuitable materials.

Wear is not an intrinsic property of the material but rather a function of the system. The elements of a wear system include the contact surfaces, lubricant, load, articulating surface speeds, motions, surface roughnesses, and temperature. The general conditions existing at the contact interface may not control wear as much as the specific load conditions existing at the asperities on the contact surfaces. Unfortunately, conditions at asperities are difficult to measure.

Wear is also a function of history. An isolated event at the beginning of a wear test can significantly change all subsequent events. If a third-body particle is caught between contact surfaces at the beginning of a test, for example, and produces a scratch on the metallic surface, the cumulative polyethylene wear at the end of the test may be 2 to 3 times what would be observed in the absence of the third-body particle.[1,2]

A wear test is a mechanical test with moving parts. Accordingly, factors such as alignment and vibration can influence the resulting wear measurement much more than other parameters. The current practice of wear testing multiple specimens in each experimental group does not necessarily resolve this issue. Only when all these factors are under reasonable control can the difference attributable to materials be measured accurately for a given design. A different degree of wear measured between two materials in an in vitro wear test does not necessarily mean the materials will perform to the same degree of difference in vivo. When the critical wear mechanisms are sufficiently simulated in vitro, then the wear test results can be claimed clinically relevant. Therefore, understanding of the wear mechanisms and the

Crystalline Region

Amorphous Region

Highly cross-linked

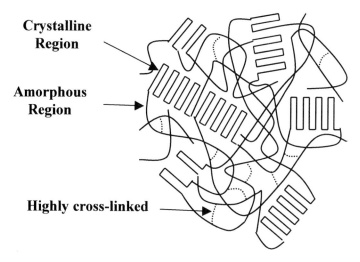

Figure 1 *A schematic diagram of the chain orientation of ultra-high molecular weight polyethylene.*

associated characteristics (surface features, wear debris size and shape) are key to successful discrimination of materials and designs. Wear mechanisms, in turn, depend on the relative material properties of the wear surfaces, the contact geometry, speeds, loads, surface roughness, and lubricant.[3]

For ultra-high molecular weight polyethylene materials, direction of motion and the frequency of motion directional changes also affect wear,[4,5] although the reason is not fully understood.[6] This type of polymer is a viscoelastic material with molecular weight that ranges from 4.5 million to 6 million. It is composed of folded chains in a crystalline arrangement and loose amorphous chains with particle domains fused together by melting (Fig. 1). Cross-linking of chains occurs only in the amorphous regions where polymer chains are free to bond with one another. A folded chain has a specific orientation. Collectively, the chains tend to align and orient themselves in a common direction when the surface is subjected to a directional force such as unidirectional rubbing. Orientation of the surface crystalline domain chains occurs easily when the surface is cut, polished, or rubbed.

Soft x-ray absorption spectroscopy[7,8] can measure the molecular orientation at polyethylene surfaces subjected to different motions of wear from a wear test and a joint simulator test. Similar results have also been observed on retrieved implants. Figure 2 shows the partial electron yield of a sample subjected to soft x-ray beams under two different sample orientations (electric field parallel to and perpendicular to the wear direction). The spectrum difference measures the degree of polymer chain alignment at the rubbed surface. Figure 3 portrays the difference spectra (subtracting the spectrum perpendicular to the electric field from the parallel spectrum), in which a larger difference indicates a higher degree of chain alignment.[9]

Table 1 summarizes the molecular orientation results of various cross-linked samples from a knee simulator test. Some degree of molecular orien-

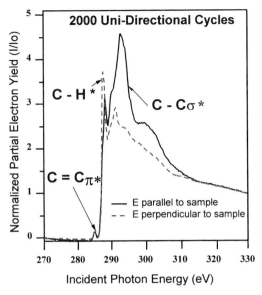

Figure 2 *Soft x-ray absorption spectra showing the molecular orientation of surface layer after 2000 unidirectional sliding cycles. E = electric field*

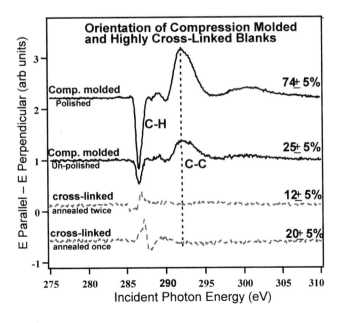

Figure 3 *Soft x-ray absorption spectra showing the effect of annealing on two different polyethylene samples. E = electric field*

Table 1 Molecular orientation of polyethylene before and after wear for different degrees of cross-linking

Sample ID	Sample	C-C (%) E ‖ Wear Direction (± 5%)	% change Between Worn-unworn (± 5%)
D10 100 kGy in N$_2$	Control No Wear	16.5	34.7
	Wear (knee sim.)	18.4	
D7 75 kGy in N$_2$	Control No Wear	4.6	74.0
	Wear (knee sim.)	18.5	
D5 50 kGy in N$_2$	Control No Wear	15.1	32.0
	Wear (knee sim.)	22.4	
D3 30 kGy in N$_2$	Control No Wear	24.4	22.0
	Wear (knee sim.)	31.3	

tation is observed for all these cases but the rate of orientation varies for different degrees of cross-linking. This can be explained by the different degrees of freedom of movement of the folded polymer chains as a function of the degrees of constraint imposed by cross-linking the chains. For the most highly cross-linked materials, more test time is necessary to break the cross links and orient the folded chains. The chain breaking mechanism is probably a mechanical fatigue process at the molecular level. The ease with which the surface molecules can be oriented or aligned has several significant implications for the observed wear phenomena. One hypothesis is that by subjecting the material to unidirectional sliding, the surface molecules align parallel to the wear direction forming a quasi-glassy material. Wear therefore occurs by tensile cracking and brittle fracture of the hardened surface layer, producing a relatively low amount of wear. When the surface is subjected to reciprocating motions, the alignment is not as good but still occurs along the axis of the motion. More loosely bound chains are worn away, generating more wear. When the surface motion is multiaxial, the surface molecules are in a constant state of flux and all fracture modes (tensile, shear, and twist) come into play. The result is a much higher amount of wear than in the other cases.[10]

According to this hypothesis, the basic wear mechanism for polyethylene can be assumed to be molecular or nanometer in scale, with loose polymer chains sticking out of the surface that are plucked away as the metallic surface slides by. Additional wear mechanisms from abrasion (third-body effect), contact pressure-induced deformation, and elastic-plastic flow from global contact stresses are seen as well. For viscoelastic and plastic materials, the formation of surface wavelets in the direction of motion has been well documented,[11] a possible explanation for the isolated islands observed on the surfaces of retrieved implants.[12] The fibril wear particles recovered from periprosthetic tissues and from lubricating fluids in joint simulators might also be explained by the elastic-plastic deformation and the rolling of thin molecular sheets detached from the substrate by fatigue cracking.

Joint simulator tests have been developed to simulate the biomechanics of human joints in a controlled condition. Results from simulator testing can provide confirmation of the material's performance for a given geometric design under a variety of operating conditions. Both controlled wear tests and joint simulator tests have been used to establish the conditions necessary to generate the types of wear damage occurring in hip and knee joints.[13-18] For example, multidirectional joint motions are necessary to generate wear rates and wear particle morphologies that are similar to in vivo experience, even though the multidirectional motions combined with the applied loads imparted by hip joint simulators do not themselves duplicate typical joint motions that patients might impart across their joints.

To simulate a knee joint experimentally requires more complex motions than simulating a hip joint simply because of the increased degrees of freedom. Knee joint simulators have only been commercially available since 1997. There are two basic designs of knee joint simulators, differing primarily in their input signals. In one type, the operator predefines the applied loading and the motions (flexion-extension, anteroposterior translation, and internal-external rotation of the femoral component on the tibial component). Surface geometry and friction do not determine the positions of the femoral component during motion as the simulator drives the components through the predetermined motions. Valid results rely on a priori knowledge of the kinematics. The correlation of results from this type of simulator with those observed on retrieved implants has not been established.[19-21]

In the second type of knee simulator design, the inputs are forces and moments, but the simulator does not directly control the position of the femoral component relative to the tibial component.[16] Instead, the femoral and tibial components are free to move relative to one another within constraints provided by the input forces and moments, the surface geometry of each component, alignment of the components, friction between the articulating surfaces, and soft-tissue restraints (simulated by springs). The limitations of this second type of knee joint simulator are that the position of the femoral component as a function of flexion angle is not directly controlled and that a larger number of inputs must be selected for each test. Studies on this second simulator type have shown that implant design has a significant independent effect on kinematics and wear[22] and that wear debris generated during wear tests of a knee design match both the morphology and the size of the particles retrieved from the knee joints of patients with the same implant design.[23] Neither type of knee simulator, however, has been validated as demonstrative that the locations and type of wear damage observed in retrieved components can be systematically recreated for a broad range of knee designs, which may be the only current means of validation. Until methods are developed to accurately measure clinical wear rates in total knee replacements, validation of knee joint simulators on the basis of matching the clinical wear rate is impossible.

There are many variations of test protocols emphasizing different motions and the frequency of such motions as exhibited by human activities. The correlations developed between a particular test protocol and retrieval studies encompass both materials and design. Such correlations take a long time to develop unless early failures are observed. As new materials and new joint simulator test protocols continue to evolve, such correlations necessarily lag behind. Therefore, the philosophy of joint simulator testing is to overstress the materials and the design. The use of the historical database also reinforces the effectiveness of the materials and design evaluations. As such, joint simulator testing remains a main staple in today's highly sophisticated material evaluation process.

A mechanistic understanding of wear provides a framework for evaluating different materials and designs. Each bearing material has individual characteristics and properties that reflect both its initial fabrication and subsequent changes that might have occurred in service. For example, polyethylene properties and wear characteristics will depend on the starting resin and the method of fabrication (molding or extrusion), but may also be affected by subsequent sterilization procedures, oxidative degradation, or crystalline reorientation under dynamic and large joint contact loads.[24] When material combinations are used to form an artificial joint, the properties of the material pair dominate the component performance. In a long-term wear process, the difference in material properties ultimately would result in different degrees of wear. But not all material property differences would result in observable or measurable changes in wear in a short-term bench wear test, and both wear bench tests and joint simulators can be considered short-term tests. What is the probability that differences that are not appreciated in the short term will amplify themselves after years of service in the body? If a series of controlled wear experiments can be conducted to elucidate the basic wear mechanisms under different loads, speeds, and lubrication environments, a clear understanding will be provided for the material pair's behavior. Durability limits could then be defined in terms of the test variables (speed, load, and cycles to failure). Often such results can be compared to an historical database, increasing the probability that drastic mistakes can be avoided. Results from extensive simulator testing can then provide confirmation of the material pair's performance for a given geometric design under a variety of operating conditions. With a clear understanding of wear mechanisms provided by the bench wear tests, an intelligent combination of bench wear testing and joint simulator studies provides the best practice in today's environment.

Relevance

Wear and simulator testing are complicated tasks. Controlled wear testing should not be routinely done to qualify a material, but rather to elucidate wear mechanisms. Simulator tests, on the other hand, can be used to conduct accelerated protocols that replicate/simulate particularly extreme conditions, thereby establishing the limits of performance for the material. The task is complicated, however, by the lack of understanding of the basic wear mechanisms under a variety of operating conditions. For now, confidence in the interpretation of wear testing data derives from successful correlation of bench test wear surfaces with retrieved implant surfaces in terms of surface texture and wear debris size and shape.[25]

Future Directions for Research

One of the critical issues in wear and joint simulator testing is how to extrapolate short-term testing results to long-term projections. This requires a solid understanding of the relationship between material structures, properties, and wear mechanisms. A carefully designed parameter study is needed to systematically examine the influence of speed, loading cycles, and motion directions on a material's behavior and the resulting wear phenomena.

Many simulation issues are related to the polyethylene wear particle generation mechanisms under different wear testing conditions. Further

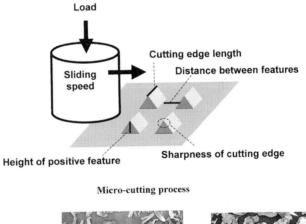

Figure 4 *Controlled wear particle generation using surface texturing and the resulting wear particles.*

research to establish the relationship between wear particle size, shape, and morphology and wear testing conditions is warranted. Sensitivity of bioactivity to particle size, shape, and morphology will provide additional parameters to engineer the polymers for longer implant life.

Because wear particle size and shape are important parameters in defining the clinical relevance of wear tests and joint simulator tests, methods to characterize different sizes and shapes are also needed. Current characterization techniques lack precision due to the lack of suitable, realistic standard reference materials for instrument calibration and testing. The technical challenge to produce well controlled wear particles for such a purpose, however, is highly complex.

Recent studies undertaken to develop controlled size and shape wear particles by rubbing ultra-high molecular weight polyethylene against a micro-textured steel surface show promise.[26-28] If successful, this concept (Fig. 4) may produce uniformly sized wear particles of various shapes. It also provides insight into the basic wear mechanisms, for example how various sizes and shapes of wear particles are produced under different wear conditions.

References

1. Barbour PS, Stone MH, Fisher J: A hip joint simulator study using new and physiologically scratched femoral heads with ultra-high molecular weight polyethylene acetabular cups. *Proc Inst Mech Eng* 2000;214:569-576.
2. McKellop H, Shen FW, DiMaio W, Lancaster JG: Wear of gamma-crosslinked polyethylene acetabular cups against roughened femoral balls. *Clin Orthop* 1999;369:73-82.
3. Liao YS, Benya PD, McKellop HA: Effect of protein lubrication on the wear properties of materials for prosthetic joints. *J Biomed Mater Res* 1999;48:465-473.
4. Firkins PJ, Tipper JL, Ingham E, Stone MH, Farrar R, Fisher J: Influence of simulator kinematics on the wear of metal-on-metal hip prostheses. *Proc Inst Mech Eng* 2001;215:119-121.
5. Saikko V, Ahlroos T: Type of motion and lubricant in wear simulation of polyethylene acetabular cup. *Proc Inst Mech Eng* 1999;213:301-310.
6. Shen MC, Hsu SM, Tesk JA, Christou A: A novel multiaxial wear tester for accelerated testing of materials. *Trans Orthop Res Soc* 1999;24:852.
7. Fischer DA, Sambasivan S, Shen MC, Hsu SM: Wear induced molecular orientation in UHMWPE measured by soft x-ray absorption. *Trans Soc Biomater* 1999;22:351.
8. Sambasivan S, Fischer DA, Shen MC, Hsu SM: Effects of wear motion on UHMWPE molecular orientation. *Trans Soc Biomater* 1999;22:194.
9. Sambasivan S, Fischer DA, Shen MC, Tesk JA, Hsu SM: Effects of annealing on UHMWPE molecular orientation, in *Transactions of the Sixth World Biomaterials Congress*. 2000, p 178.
10. Bragdon CR, O'Connor DO, Lowenstein JD, Jasty M, Syniuta WD: The importance of multidirectional motion on the wear of polyethylene. *Proc Inst Mech Eng* 1996;210:157-165.
11. Chen YM, Ives LK, Dally JW: Numerical simulation of sliding contact over a half-plane. *Wear* 1995;185:83-91.

12. Wimmer MA, Andriacchi TP, Natarajan RN, et al: A striated pattern of wear in ultra-high-molecular-weight polyethylene components of Miller-Galante total knee arthroplasty. *J Arthroplasty* 1998;13:8-16.

13. Baldini TH, Wright TM, Estupiñán JA, Bartel DL: An apparatus for studying wear damage in UHMWPE: Description and initial test results, in Yoganathan AP (ed): *1998 Advances in Bioengineering*. New York, NY, ASME, 1998, Vol 39, pp 347-348.

14. Blunn GW, Walker PS, Joshi A, Hardinge K: The dominance of cyclic sliding in producing wear in total knee replacements. *Clin Orthop* 1991;273:253-260.

15. Smith SL, Unsworth A: A five-station hip joint simulator. *Proc Inst Mech Eng* 2001;215:61-64.

16. Walker PS, Blunn GW, Broome DR, et al: A knee simulating machine for performance evaluation of total knee replacements. *J Biomechanics* 1997;30:83-89.

17. Barbour PS, Stone MH, Fisher J: A hip joint simulator study using simplified loading and motion cycles generating physiological wear paths and rates. *Proc Inst Mech Eng* 1999;213:455-467.

18. DesJardins JD, Walker PS, Haider H, Perry J: The use of a force controlled dynamic knee simulator to quantify the mechanical performance of total knee replacement designs during functional activity. *J Biomechanics* 2000;33:1231-1242.

19. Deluzio KJ, Muratoglu OK, O'Connor DO, et al: Development of an in vitro knee delamination model in a knee simulator with physiologic load and motion, in *Transactions of the Sixth World Biomaterials Congress*. 2000, p 842.

20. Ries M, Banks S, Sauer W, Anthony M: Abrasive wear simulation in total knee arthroplasty. *Trans Orthop Res Soc* 1999;45:853.

21. Tamura J, Clarke I, Kawanabe K, Schroeder D, Gufstason A: Tibial polyethylene micro-wear phenomena related to TKR kinematics. *Trans Orthop Res Soc* 2000;46:433.

22. DesJardins JD, Walker PS, Haider H, Perry J: The use of a force controlled dynamic knee simulator to quantify the mechanical performance of total knee replacement designs during functional activity. *J Biomechanics* 2000;33:1231-1242.

23. Beaule P, Campbell P, Walker PS, Schmalzreid TP, Dorey F, Blunn GW: Characterization of polyethylene wear particles retrieved from tissues and knee simulator lubricants. *Trans Soc Biomater* 2000;23:58.

24. McKellop HA, Shen FW, Campbell P, Ota T: Effect of molecular weight, calcium stearate, and sterilization methods on the wear of ultra high molecular weight polyethylene acetabular cups in a hip joint simulator. *J Orthop Res* 1999;17:329-339.

25. Shen MC, Hsu SM, Tesk JA, Christou A: Wear analysis of UHMWPE using a load sum method, in *Transactions of the Sixth World Biomaterials Congress*. 2000, p 174.

26. Fang H-W, Shen MC, Cho U, Tesk JA, Christou A, Hsu SM: Generation of different UHMWPE particle shape by wear through surface texturing. *Trans Soc Biomater* 1999;22:506.

27. Fang H-W, Sengers JV, Hsu SM: Controlled size and shape UHMWPE wear particle generation using a silicon micro-fabricated surface texturing technique. *Trans Soc Biomater* 2001;24:30.

28. Fang H-W, Sengers JV, Hsu SM: Generation of controlled UHMWPE wear particle size and shape by surface texturing: Relationship between surface feature dimensions to particle size and shape. *Trans Soc Biomater* 2001;24:585.

What design factors influence wear behavior in total knee replacement?

Design factors that influence the durability of total knee components and, therefore, the generation of wear debris include articular geometry, component thickness, and material and surface characteristics of the components.[1-17] The influence of these factors can be considered in relation to the other factors that influence wear. Load, for example, has a direct influence on wear and is primarily a patient factor; however, design can affect the quadriceps muscle lever arm (and the resulting moment generated by the quadriceps) by the dishing of both the tibial articular surface and the patellar flange of the femoral component. Wear must be evaluated, therefore, under loading conditions that represent normal walking gait[15-16] as well as other activities of daily living, such as stair ascent (Table 1).[17-19]

Sliding distance is another factor that is primarily influenced by the patient through activity level but is also controlled by design of the articular surfaces. In fact, bearing surface design, particularly the femorotibial constraint, plays a major role in determining sliding distance.[20-24] Material factors can significantly affect the type of designs possible, particularly with regard to conformity.[25-31] The type of motion between the bearing surfaces, likewise influenced by patient and design factors, also affects wear. Sliding, pure rolling (no shear transmitted between the surfaces), tractive rolling (shear transmitted), and twisting may create different types of surface damage and wear.[32,33]

Wear can occur at a steady state over time through adhesive and abrasive mechanisms or can be time-dependent, eg, delamination occurring only after the accumulation of many cycles of fatigue-related damage. Steady state

Table 1 Typical tibial-femoral loading conditions*

Knee flexion angle	Assumed multiple of body weight	Applied load	Activity	Loading conditions
0°	4.0	2903 N	Walking	High cycle, low load
60°	5.0	3629 N	Stair ascent	Low cycle, high load

*For 60-year-old male, weight 726 N, height 1.73 m.[15-19]

Adapted with permission from Heim CS, Postak PD, Greenwald AS: Factors influencing the longevity of UHMWPE tibial components. Instr Course Lect 1996;45:303-312.

wear may be roughly linear with sliding distance (number of cycles). Time-dependent wear is more complex to analyze and predict.[21]

Fixed and Mobile Bearing Designs

For fixed bearing total knee replacements (TKRs), kinematics is controlled by bearing surface geometry in both the frontal and sagittal planes. Frontal geometry affects internal-external rotation. More dished and conforming surfaces (such as the original Total Condylar design) are more constrained than shallower surfaces (eg, the Miller/Galante). Sagittal geometry affects both internal-external rotation and AP displacement. The most influential parameter is the sagittal radius of the tibial surface. In general, the less the constraint between the bearing surfaces, the greater the relative sliding and the greater the wear. Rolling and tractive shear occur in all TKR bearings due to imposed forces and to friction.[34] Variations in the magnitudes of sliding and rotation as a function of design have been demonstrated in both laboratory knee simulator experiments and in vivo fluoroscopy studies.[20,24,35-37]

Although extensive work has been done in measuring and predicting contact areas in TKRs, no clear relationship between wear and contact area has been demonstrated when other important variables (load and sliding distance) are held constant.[38] The primary factor is that small contact areas occur in designs with low constraint, which leads to greater sliding and increased wear.[34] For example, delamination wear is not necessarily produced by the least contact area, but rather by a preponderance of sliding.[21]

An additional factor in fixed bearing TKRs is backside wear between the polyethylene tibial insert and the metallic tibial tray.[39] Backside wear is a function of the fit between the insert and tray, which is in turn influenced both by design and by manufacturing tolerances, the roughness of the surface finish (especially the metallic surface), and the material from which the metallic tray is fabricated.

In contrast to fixed bearing TKRs, kinematics in mobile bearing TKRs is controlled by the geometry of both the upper (articular) and lower (mobile) bearing surfaces, as well as the design of the connection between the mobile polyethylene insert and the metallic tray. Upper surface geometry can be fully conforming (as in the Oxford, MBK, and Rotaglide designs) and thus restricted to uniaxial motion with complete contact.[40-42] The upper bearing surface can also be a combination of fully and partially conforming at different ranges of flexion (as in the LCS design); a reduced contact area with flexion allows some rolling and sliding.[43]

The lower bearing surface can be unicompartmental with a small surface area (as in the Oxford design) or bicompartmental with a larger surface area (such as the LCS, Rotaglide, and MBK designs). The mobile bearing insert-tray connection can be designed to allow primarily AP translation (eg, the Oxford), only internal-external rotation (eg, LCS), or both (eg, MBK and Rotaglide). Surprisingly, in vivo fluoroscopic studies show that relative displacements and rotations in mobile bearing designs are similar to those in less constrained fixed bearing designs,[36,37] a finding also observed in force-input types of knee simulator studies.[20,41] The probable explanation is that under large compressive joint loads, the friction coefficient between the bearing surfaces is sufficient to limit or even prevent relative sliding.[34]

On the upper articular bearing surface, full conformity is likely to have less wear than partial conformity, as observed on retrieved components of the Oxford design[40] and in knee simulator studies of the MBK design[41] where only minor adhesive wear occurred. On the lower mobile bearing surface, restricting the motion to rotation is likely to produce less wear than allowing both rotation and sliding motions. Larger surface areas at this bearing in total condylar designs appear to contribute to abrasive, third-body wear from entrapped debris and lack of lubrication.[38] Scratching of the metallic tray surface and greater polyethylene wear from this bearing than from the upper articular surface support this hypothesis.

It appears that wear can be reduced in mobile bearing designs compared with fixed bearing designs, primarily due to the larger contact areas. However, the wide design variations in both fixed bearing and mobile bearing categories of TKR preclude any generalization.

Considerations of Contact Stresses and Areas

Although a direct relationship between contact areas, contact stresses, and wear mechanisms remains elusive, experimental measurements of contact (for example, using statically applied loads and pressure sensitive film) are routinely used to establish the influence of design factors (Fig. 1). While

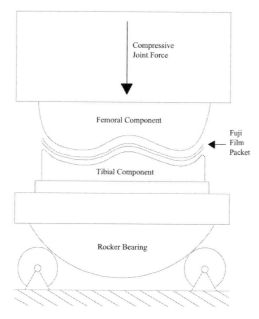

Figure 1 Contact area and surface stress distribution test apparatus. *(Adapted with permission from Heim CS, Postak PD, Greenwald AS: Factors influencing the longevity of UHMW-PE tibial components. Arch Am Acad Orthop Surg 1997;1:51-56.)*

kinematics also play an important role, greater contact stresses are assumed to cause more wear based on the hypothesis that pitting and delamination in polyethylene tibial bearing surfaces[44] are linked to subsurface fatigue failures.[45-49] In the studies described in the following paragraphs, differences in UHMWPE stresses have been normalized to their respective tensile yield strengths to facilitate comparison. The potential for wear damage can be determined by analyzing the magnitude of the surface stresses concomitant with their respective contact areas. Thus, high contact stresses acting over a large contact area have the greatest propensity for early fatigue failure.

Tibial-femoral condylar conformity influences the amount of the contact surface within which the polyethylene would be expected to exceed its yield strength. A comparison of two commercial designs that differ only in the articular geometry of the tibial component shows that contact areas and surface stresses are position-dependent (Fig. 2). In one design (TC), the tibial insert has a flat, minimally constraining geometry; the other insert (LSI) is intended to provide joint stability through an increase in surface curvature. The more curved design would appear to have less potential for polymer damage at full extension due to the small amount of post-yield contact area, while the less constrained design has less potential for damage at 60° of flexion because the conformity of the LSI design decreases more rapidly than the TC design.

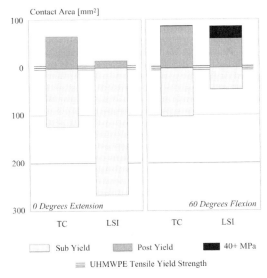

Figure 2 *Contact areas by surface stress range for comparison of tibial-femoral conformal geometries (n = 3 for each condition). Advantim Modular Total Condylar (TC) and Ligament Supplementing (LSI) Total Knee Designs (Wright Medical Technology Inc, Arlington, TN) were evaluated using 12-mm tibial components sterilized by gamma irradiation. The overall bar height depicts the total contact area. Total post-yield contact area is represented by the height of the bar above the tensile yield strength inclusive of the 40+ MPa area. (Reproduced with permission from Heim CS, Postak PD, Greenwald AS: Factors influencing the longevity of UHMWPE tibial components. Arch Am Acad Orthop Surg 1997;1:51-56.)*

Femoral component finish also influences surface stress distributions. Belt finishing is a manual process that uses abrasive belts to rough-in articulating surfaces of cast femoral components, while precision ground components are shaped via a computer-aided grinder. When the two techniques were compared, precision grinding demonstrated a lower potential for polyethylene damage with a marked decrease in the post-yield contact area (Fig. 3). The more precise technique maintains better femoral component tolerances, thus maintaining the conformity designed into the components. As design specifications continue to call for higher required tolerances, precision manufacturing techniques should become the industry standard.

The influence of polyethylene properties can be examined using the same approach. One study compared conventional polyethylene to an enhanced polyethylene, keeping design and thickness factors constant. The enhanced material (Hylamer-M) was intended to possess higher tensile yield strength, increased crack propagation resistance, and decreased creep deformation.[50] The enhanced material, however, also had a larger elastic modulus, which resulted in a reduction in overall contact area (Fig. 4). These findings are descriptive of increased surface stresses, which the designers hoped would be offset by enhanced material properties. In short-term clinical use, the enhanced material gave rise to fatigue-induced component failure.

Properties of polyethylene that has been sterilized by gamma irradiation in air change with shelf life. Chain scission by the radiation allows oxidation, causing embrittlement and reduced wear resistance.[31,51-62] The degradation is especially detrimental because it occurs primarily near the surface of the material.[61-63] When polyethylene inserts were gamma irradiated in air and stored for 5 years prior to testing, contact mechanics were found to be considerably more damaging than identical components stored for only 3

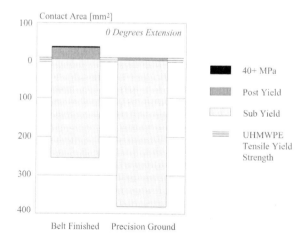

Figure 3 *Contact areas by surface stress range for comparison of femoral components prepared with two different finishing techniques (n = 3 for each finishing type). Apollo Total Knee Design (Sulzer Orthopedics Inc, Austin, TX). (Reproduced with permission from Heim CS, Postak PD, Greenwald AS: Factors influencing the longevity of UHMWPE tibial components. Arch Am Acad Orthop Surg 1997;1:51-56.)*

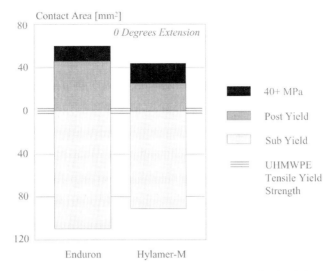

Figure 4 *Contact areas by surface stress range for comparison of polyethylene properties in the same knee design (n = 3 for each polyethylene type). AMK Standard Total Knee Design (DePuy Orthopaedics Inc, Warsaw, IN). (Reproduced with permission from Heim CS, Postak PD, Greenwald AS: Factors influencing the longevity of UHMWPE tibial components. Arch Am Acad Orthop Surg 1997;1:51-56.)*

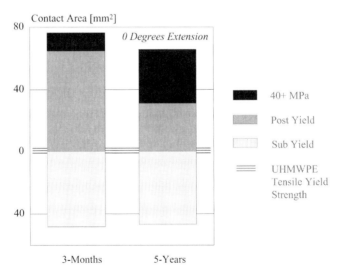

Figure 5 *Contact areas by surface stress range for polyethylenes gamma irradiated in air and stored for 3 months or 5 years in the same knee design (n = 3 for each combination of polyethylene type and shelf life). Performance PCL Retaining Total Knee Design (Kirschner Medical Corporation, Timonium, MD). (Reproduced with permission from Heim CS, Postak PD, Greenwald AS: Factors influencing the longevity of UHMWPE tibial components. Arch Am Acad Orthop Surg 1997;1:51-56.)*

161

Table 2 Patient data grouped by implant shelf life

Shelf life prior to implantation	Number implanted	Average patient age (years)	Average patient weight (kg)	Average component shelf life (years)	Survival after 5 years (%)
0 – 4 years	93	69	85.9	1.9	100.0
> 4 – 8 years	21	68	98.3	5.9	88.6
> 8 – 11 years	21	65	95.5	9.2	79.2
Total	135	68	89.3	3.6	95.1

Reproduced with permission from Bohl JR, Bohl WR, Postak PD, Greenwald AS: The effects if shelf life on clinical outcome for gamma sterilized polyethylene tibial components. Clin Orthop 1999;367:28-38.

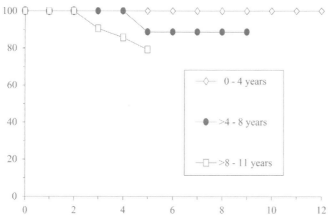

Figure 6 *Survival data based on implant shelf life. X axis, years in vivo; Y axis, survival rate. (Reproduced with permission from Bohl JR, Bohl WR, Postak PD, Greenwald AS: The effects of shelf life on clinical outcome for gamma sterilized polyethylene tibial components. Clin Orthop 1999;367:28-38.)*

months (Fig. 5). Degradation causes an increase in density and a subsequent increase in elastic modulus that adversely affects the contact stresses.

Clinical studies substantiate these in vitro results,[64] identifying component shelf life as a major contributor to wear-related failure. For example, between 1985 and 1994, 135 Synatomic knee components were implanted by a single surgeon into 105 patients. All implants were packaged in an air environment and sterilized by gamma irradiation. Implants were grouped by length of shelf life (Table 2) and survival rates were determined for each group (Fig. 6). All failures exhibited extreme polyethylene degradation (Fig. 7). These data suggest that an expiration date of 5 years is appropriate for devices gamma irradiated in air, supporting the recent safety notice from the United Kingdom Medical Devices Agency calling for the removal from inventory of any UHMWPE component that is more than 5 years beyond its date of manufacture and sterilization.[65]

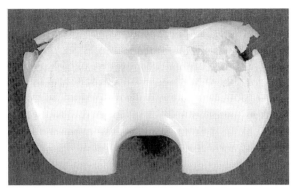

Figure 7 *This component, with a 9.6-year shelf life, was retrieved 13 months after implantation. (Reproduced with permission from Bohl JR, Bohl WR, Postak PD, Greenwald AS: The effects of shelf life on clinical outcome for gamma sterilized polyethylene tibial components. Clin Orthop 1999;367:28-38.)*

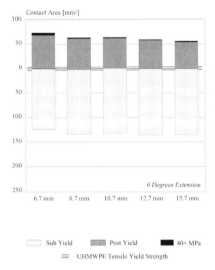

Figure 8 *Contact areas by surface stress range for comparison of tibial component thickness (n = 5 for each tibial insert thickness). Genesis II Cruciate Retaining Total Knee Design (Smith & Nephew Inc, Memphis, TN). (Reproduced with permission from Heim CS, Postak PD, Greenwald AS: Factors influencing the longevity of UHMWPE tibial components. Arch Am Acad Orthop Surg 1997;1:51-56.)*

Static loading and pressure measurements have also been used to examine the influence of tibial plateau thickness on surface stress distributions. Analytical results would suggest that higher conformity and thicker polyethylene components result in decreased contact stresses. Experimentally, however, no dramatic increase in the potential for polyethylene damage was observed as insert thickness was decreased (Fig. 8), suggesting that assumptions in the original analyses (such as linearly elastic polyethylene material behavior) limit their usefulness in predicting the influence of thickness.

Knee Joint Simulators

It is impossible to predict wear in a new design from geometrical parameters alone because numerous patient and surgical variables, many of which interact with design, also affect wear. Similarly, contact areas and stresses, which can also be determined from numerical analyses, remain as yet unreliable tools for predicting wear. Knee simulator testing may become necessary in order to compare a new design with a clinically evaluated standard design.

The predictive power of knee simulator testing is still uncertain. Several simulator designs and testing methods are currently under investigation.[24,31] Force-input knee simulators (such as the Stanmore-Instron system) allow individual TKR designs to produce their own kinematics during the test, which results in reliable comparative data. This type of system, however, requires selection of a number of test variables, many of which are not well defined in terms of either appropriate standard values or their impact on the resulting wear created in the machine.[24] Displacement-input simulators (such as the MTS and AMTI machines) require input of kinematics data. Because kinematics are unknown during the preclinical phase of design (ie, no devices have been implanted from which kinematics data might be obtained through fluoroscopy), the clinical relevance of the data from such systems is unclear.[66]

Relevance

The integrity of a polyethylene bearing surface does not rest primarily with the sizes of the contact areas but with the magnitudes and types of surface and subsurface stresses within these regions. The principle of the hypothesis relating surface stress to wear is that large areas of high surface stress have the most potential for wear damage.[1] Although this relationship has not been clearly defined for the nonconforming articulations that exist in TKR designs, it can be used to suggest the influence of important design factors on wear performance. The static measurement of contact areas and pressures also has limitations, failing to reflect the cyclic nature of activities of daily living (Table 1).

The lack of understanding of wear mechanisms in TKRs leaves many unanswered questions. Which material properties control wear resistance? How do these properties change with processing (eg, sterilization)? How do they change with the gross permanent deformation that occurs as a result of high cyclic contact stresses and variations in articular kinematics (eg, with variations in design, surgical alignment, and soft-tissue balancing)?

Design considerations affect both the volumetric wear rate and the morphology of the wear particles that are produced in TKRs. Because long-term failure can result from the accumulation of wear debris, determination of the wear behavior of fixed versus mobile bearing designs is important. Design factors that minimize wear without adversely affecting function (or certainly while improving function) should be incorporated into modern designs.

Future Directions for Research

Understanding the precise mechanisms of wear and the material properties that control them will afford opportunities to tailor designs and materials to ensure long-term wear resistance. Considerable circumstantial evidence exists that fatigue plays a significant role in the wear mechanisms of TKRs. Accordingly, research should be directed into polyethylene fracture mechanics (the initiation and propagation of cracks) as a function of property variations, mixed modes of loading, and degradation. Another focus would be analytical efforts to develop polyethylene failure criteria based on stress, strain, or energy concepts.

The introduction of elevated cross-linked polyethylenes into TKRs certainly merits attention. The cross-linking processes change the chemical structure of the polyethylene and affect both static and fatigue properties, including a decreased resistance to fatigue crack propagation[67] that might make them more susceptible to wear in total knee applications.

More research is warranted into the wear rates of mobile bearing designs and the influences of many of the same factors found to affect wear in fixed bearing knees, such as bearing surface geometry, component thickness, and surface finish.[36,68]

A better understanding of the impact of manufacturing techniques (casting, grinding, and polishing) on component dimensions could lead to the development of accepted standards for many of these manufacturing steps.

Backside wear between the inferior surface of the polyethylene insert and the superior surface of the metallic tray remains controversial.[69,70] Resistance to backside wear depends on the integrity of the locking mechanism between insert and tray. Clinical and laboratory research into the role of backside wear in total knee failure is needed to determine the scope of the problem and potential solutions.

The importance of wear evaluations for TKRs and the potential for knee simulator studies to provide useful preclinical wear evaluations underscore the importance of further research to standardize simulator test protocols and establish performance of contemporary designs for which wear and clinical behavior are well known. Methodical testing of the effect of geometric design variables could then be performed with confidence, or knee simulator machines could be redesigned to better reproduce knee kinematics and loading conditions[71] and wear patterns and particles observed in retrieved components.[72]

References

1. Bartel DL, Bicknell VL, Wright TM: The effect of conformity, thickness, and material on stresses in ultra-high molecular weight components for total joint replacement. *J Bone Joint Surg Am* 1986;68:1041-1051.
2. Bartel DL, Burstein AH, Toda MD: The effects of conformity and plastic thickness on contact stress in metal-backed plastic implants. *J Biomech Eng* 1985;107:193-199.
3. Collier JP, Mayor MB, Suprenant VA, Suprenant HP, Dauphinais LA, Jensen RE: The biomechanical problems of polyethylene as a bearing surface. *Clin Orthop* 1990;261:107-113.

4. Kurtz SM, Bartel DL, Rimnac CM: Post-irradiation aging and the stresses in UHMWPE components for total joint replacement. *Trans Orthop Res Soc* 1994;19:584.

5. Lotke PA, Ecker ML: Influence of positioning of prosthesis in total knee replacement. *J Bone Joint Surg Am* 1977;59:77-79.

6. Mirra JM, Marder RA, Amstutz HA: The pathology of failed total joint arthroplasty. *Clin Orthop* 1982;170:175-183.

7. Moreland JR: Mechanisms of failure in total knee arthroplasty. *Clin Orthop* 1988;226:49-64.

8. Murase K, Crowninshield RD, Pederson DR, Chang TS: An analysis of tibial component design in total knee arthroplasty. *J Biomech* 1983;16:13-22.

9. Rose RM, Ries MD, Paul IL, Crugnola AM, Ellis E: On the true wear rate of ultra-high molecular weight polyethylene in the total knee prosthesis. *J Biomed Mater Res* 1984;18:207.

10. Seedhom BB, Wallbridge NC: Walking activities and wear of prostheses. *Ann Rheum Dis* 1985;44:838.

11. Soudry M, Walker PS, Reilly DT, Kurosawa H, Sledge CB: Effects of total knee replacement design on femoral-tibial contact conditions. *J Arthroplasty* 1986;1:35-45.

12. Walker PS, Hsieh HH: Conformity in condylar replacement knee prostheses. *J Bone Joint Surg Br* 1977;59:222-228.

13. Walker PS: Bearing surface design in total knee replacement. *Eng Med* 1988;17:194.

14. Windsor RE, Scuderi GR, Moran MC, Insall JN: Mechanisms of failure of the femoral and tibial components in total knee arthroplasty. *Clin Orthop* 1989;248:15-19.

15. Morrison JB: The mechanics of the knee joint in relation to normal walking. *J Biomech* 1970;3:51-61.

16. Kadaba MP, Ramakrishnan HK, Wootten ME: Measurement of lower extremity kinematics during level walking. *J Orthop Res* 1990;8:383-392.

17. Andriacchi TP, Andersson GBJ, Fermier RW, Stern D, Galante JO: A study of lower-limb mechanics during stair-climbing. *J Bone Joint Surg Am* 1981;62:749-757.

18. Morrison JB: Function of the knee joint in various activities. *Biomed Mater Eng* 1969;4:573-580.

19. Diem K, Lentner C: *Scientific Tables*. Switzerland, Ciba Geigy Limited, 1973, p 711.

20. DesJardins JD, Walker PS, Haider H, Perry J: The use of a force-controlled dynamic knee simulator to quantify the mechanical performance of total knee replacement designs. *J Biomechanics* 2000;33:1231-1242.

21. Sathasivam S, Walker PS: A computer model to predict subsurface damage in tibial inserts of total knees. *J Orthop Res* 1998;16:564-571.

22. Sathasivam S, Walker PS: The conflicting requirements of laxity and conformity in total knee replacement. *J Biomechanics* 1999;32:239-247.

23. Walker PS, Blunn GW, Broome DR, et al: A knee simulating machine for performance evaluation of total knee replacements. *J Biomechanics* 1997;30:83-89.

24. Walker PS, Blunn GW, Perry JP, et al: Methodology for long-term wear testing of total knee replacements. *Clin Orthop* 2000;372:290-301.

25. Bell C, Walker PS, Abeysundera M, Simmons J, King P, Blunn GW: Effect of oxidation on delamination of ultra-high-molecular-weight polyethylene tibial components. *J Arthroplasty* 1998;13:280-290.

26. Bell CJ, Blunn GW, Walker PS, Bosman H, Bartlett G, Scott R: Oxidation and wear resistance of directly moulded UHMWPE. *Trans Orthop Res Soc* 1999;24:97.

27. Blunn GW, Joshi AB, Lilley PA, et al: Polyethylene wear in unicondylar knee prostheses: 106 retrieved Marmor, PCA and St. Georg tibial components compared. *Acta Orthop Scand* 1992;63:247-255.

28. Blunn GW, Joshi AB, Minns RJ, et al: Wear in retrieved condylar knee arthroplasties. *J Arthroplasty* 1997;12:281-290.

29. Collier JP, Sperling DK, Currier JH, Sutula LC, Saum KA, Mayor MB: Impact of gamma sterilization on clinical performance of polyethylene in the knee. *J Arthroplasty* 1996;11:377-389.

30. Kurtz SM, Rimnac CM, Bartel DL: Degradation rate of ultra-high molecular weight polyethylene. *J Orthop Res* 1997;15:57-61.

31. Rimnac CM, Klein RW, Betts F, Wright TM: Post-irradiation aging of ultra-high molecular weight polyethylene. *J Bone Joint Surg Am* 1994;76:1052-1056.

32. Estupinan JA, Bartel DL, Wright TM: Residual stresses in ultra-high molecular weight polyethylene loaded cyclically by a rigid moving indenter in nonconforming geometries. *J Orthop Res* 1998;16:80-88.

33. Wimmer MA, Andriacchi TP, Natarajan RN: A striated pattern of wear in ultra-high molecular weight polyethylene components of Miller-Galante total knee arthroplasty. *J Arthroplasty* 1998;13:8-16.

34. Sathasivam S, Walker PS: A computer model with surface friction for the prediction of total knee kinematics. *J Biomechanics* 1997;30:177-184.

35. Dennis DA, Komistek RD, Hoff WA, Gabriel SM: In vivo knee kinematics derived using an inverse perspective technique. *Clin Orthop* 1996;331:107-117.

36. Stiehl JB, Dennis DA, Komistek RD, Keblish PA: In vivo kinematic analysis of a mobile bearing total knee prosthesis. *Clin Orthop* 1997;345:60-66.

37. Stiehl JB, Komistek RD, Dennis DA: In vivo kinematic comparison of posterior cruciate retention or sacrifice with a mobile bearing total knee arthroplasty. *Trans Orthop Res Soc* 1997;22:658.

38. Sathasivam S, Walker PS: Will mobile bearing knees reduce polyethylene wear compared to fixed bearing designs? *Trans Orthop Res Soc* 1999;24:961.

39. Engh GA, Rao A, Ammeen DJ, Sychterz CJ: Tibial baseplate wear: A major source of wear debris with contemporary modular knee implants. Presented at the 67th Annual Meeting of the American Academy of Orthopaedic Surgeons, SE037. Orlando, FL, March 2000.

40. Psychoyios V, Crawford RW, O'Connor JJ, Murray DW: Wear of conguent meniscal bearings in unicompartmental knee arthroplasty. *J Bone Joint Surg Br* 1998;80:976-982.

41. Bell CJ, Walker PS, Sathasivam S, Campbell PA, Blunn GW: Differences in wear between fixed bearing and mobile bearing knees. *Trans Orthop Res Soc* 1999;24:962.

42. Polyzoides AJ, Dendrinos GK, Tsakonas H: The Rotaglide total knee arthroplasty. *J Arthroplasty* 1996;11:453-459.

43. Buechel FF, Keblish PA, Lee JM, Pappas MJ: Low contact stress meniscal bearing unicompartmental knee replacement: Long-term evaluation of cemented and cementless results. *J Orthop Rheumatol* 1994;7:31-41.

44. Collier DE, Burstein AH, Fenning JB, Keblish P Jr, Buechel FF, Walker PS: The relationship of polyethylene wear to the failure of total joint replacements. *Contemp Orthop* 1987;15:6.

45. Connelly GM, Rimnac CM, Wright TM, Hertzberg RW, Manson JA: Fatigue crack propagation behavior of ultrahigh molecular weight polyethylene. *J Orthop Res* 1984;2:119-125.

46. Elbert KE, Wright TM, Rimnac CM, et al: Fatigue crack propagation behavior of ultra high molecular weight polyethylene under mixed mode conditions. *J Biomed Mat Res* 1994;28:181-187.

47. Mowery C, Botte M, Bradley G: Fracture of polyethylene tibial components in a total knee replacement: A case report. *Orthopedics* 1987;10:309-313.

48. Pruitt L, Koo J, Rimnac CM, Suresh S, Wright T: Cyclic compressive loading results in fatigue cracks in ultra high molecular weight polyethylene. *J Orthop Res* 1995;13:143-146.

49. Weightman B, Isherwood DP, Swanson SAV: The fracture of ultrahigh molecular weight polyethylene in the human body. *J Biomed Mater Res* 1979;12:669.

50. Champion AR, Li S, Saum K, Howard E, Simmons W: The effect of crystallinity on the physical properties of UHMWPE. *Trans Orthop Res Soc* 1994;19:585.

51. Rose RM, Goldfarb EY, Ellis EJ, Crugnola A: Radiation sterilization and the wear rate of polyethylene. *J Orthop Res* 1984;2:393-400.

52. Sauer WL, Weaver KD, Beals NB: Fatigue performance of UHMWPE orthopaedic bearing material. *Trans Soc Biomater* 1994;17:122.

53. Birkinshaw C, Buggy M, O'Neill M: The effect of gamma radiation on the physical structure and mechanical properties of ultrahigh molecular weight polyethylene. J *Appl Polymer Sci* 1989;38:1967-1973.

54. Birkinshaw C, Buggy M, O'Neill M: Mechanism of aging in irradiated polymers. *Polymer Degrad Stabil* 1988;22:285-294.

55. Bostrom MPG, Bennett AP, Rimnac CM, Wright TM: Degradation in polyethylene as a result of sterilization, shelf storage and in vivo use. *Trans Orthop Res Soc* 1994;19:288.

56. Hawkins ME, Gsell R: Effect of nitrogen on radiation-induced oxidation to ultra-high molecular-weight polyethylene (UHMWPE). Trans Soc Biomater 1994;17:181.

57. Kurth M, Eyerer P, Cui D: Effects of radiation sterilization on UHMW-polyethylene, in *Proceedings of ANTEC.* 1987, p 1193.

58. Li S, Saum KA, Collier JP, Kasprazak D: Oxidation of UHMWPE over long time periods. *Trans Soc Biomater* 1994:425.

59. Lue CT, Ellis EJ, Crugnola A: Effects of gamma irradiation on ultra-high-molecular-weight polyethylene, in *Proceedings of SPE/ANTEC.* 1981, p 246.

60. Salorey S, Shinde A: Irradiation of ultra high molecular weight polyethylene. *Polymer Preprints* 1985;26:118-119.

61. Saum KA: Oxidation vs. depth and time for polyethylene gamma sterilized in air. *Trans Orthop Res Soc* 1994;19:174.

62. Kamel I, Finegold L: A model for radiation-induced changes in ultrahigh-molecular-weight polyethylene. *J Polymer Sci Polymer Physics* 1985;23:2407.

63. Elbert KE, Kurth M, Bartel DL, Eyerer P, Rimnac CM, Wright TM: In vivo changes in material properties of polyethylene and their effects on stresses associated with surface damage of polyethylene components. *Trans Orthop Res Soc* 1988;13:53.

64. Bohl JR, Bohl WR, Postak PD, Greenwald AS: The effects of shelf life on clinical outcome for gamma sterilized polyethylene tibial components. *Clin Orthop* 1999;367:28-38.

65. Great Britain, Medical Devices Agency: Ultra-high molecular weight polyethylene (UHMWPE) components of joint replacement implants. London, Department of Health (Safety Notice 9816), 1998.

66. Burgess IC, Kolar M, Cunningham JL, Unsworth A: Development of a six station knee wear simulator and preliminary wear results. *Proc Inst Mech Engrs* 1997;211:37-47.

67. Baker DA, Hastings RS, Pruitt L: Study of fatigue resistance of chemical and radiation crosslinked medical grade ultrahigh molecular weight polyethylene. *J Biomed Mater Res* 1999;46:573-581.

68. Hartford JM, Banit D, Hall K, Kaufer H: Radiographic analysis of low contact stress meniscal bearing total knee replacements. *J Bone Joint Surg Am* 2001;83:229-234.

69. Parks NL, Engh GA, Topoleski LD, Emperado J: Modular tibial insert micromotion: A concern with contemporary knee implants. *Clin Orthop* 1998;356:10-15.

70. Wasielewski RC, Parks N, Williams I, Suprenant H, Collier JP, Engh G: Tibial insert undersurface as a contributing source of polyethylene wear debris. *Clin Orthop* 1997;345:53-59.
71. Taylor S, Walker PS, Perry J, Cannon SR, Woledge R: The forces in the distal femur and the knee during walking and other activities measured by telemetry. *J Arthroplasty* 1998;13:428-437.
72. Peterson C, Benjamin JB, Szivek JA, Anderson PL, Shriki J. Wong M: Polyethylene particle morphology in synovial fluid of failed knee arthroplasty. *Clin Orthop* 1999;359:167-175.

What design factors influence wear behavior at the articulating surfaces in total hip replacement?

Although clinical series have only recently demonstrated a formal relationship,[1] the causal linkage between wear debris burden and osteolytic aseptic loosening in total hip arthroplasty has been well recognized for nearly two decades.[2] One of the primary approaches to the osteolysis problem is to address ways to reduce the root cause—wear debris; one approach to reduction of wear debris is improvement of implant designs.

Mechanical design attributes cannot be uncoupled from materials, and the menu of available bearing surface combinations is diverse. In addition to metal on traditional polyethylene, accepted bearing surface pairs include ceramic on polyethylene, ceramic on ceramic, second generation metal-on-metal, and metal on elevated cross-linked polyethylene. The latter four bearing surface combinations outperform metal on traditional polyethylene in terms of volumetric wear rate, although various ancillary issues (eg, absolute particle numbers and relative osteolytic potency) currently confound direct comparison. Because their wear rates are so low and therefore difficult to measure, relatively little is known about how these four couples are influenced by mechanical design parameters. By contrast, a wealth of information exists on metal-on-polyethylene designs, which comprise the vast majority of implants currently in service.

For the abrasive and adhesive wear mechanisms widely held to engender the micron- or submicron-sized polyethylene particles responsible for osteolysis, debris removal from the bearing surface depends on the combined interaction of contact stress, articulation kinematics, and the tribologic properties of the bearing surface couple. At least for the idealization of uniformly loaded isotropic surfaces with nominal contact stresses in the linearly elastic regime, that combined interaction can be described by the classic Archard relationship.[3] This relationship holds that wear depth is simply the linear product of contact stress, sliding distance, and a (constant) surface-dependent wear coefficient. Although the actual situation in total hip arthroplasty of course differs substantially from that idealization, the systematic control of these three primary variables has been the conceptual basis for laboratory hip wear simulators.[4] Bringing these primary variables ever more closely in line with clinical reality has continued to drive advances in laboratory simulator technology.[5] Although laboratory simulators have been most useful in screening bearing surface couples (eg, materials, manufacturing processes, surface finishes, and sterilization parameters), they have also been helpful in understanding the effects of mechanical design variables,[6] such as femoral head size, head-cup congruency, polyethylene liner geometry, and screw-hole placement.

Another vehicle for assessing the wear effects of mechanical design variables has been the use of total hip-specific versions of the Archard relationship,[7,8] an approach that lends itself well to parametric studies of individual geometric factors. In combination, laboratory simulators and computational models have provided compelling evidence that bearing surface volumetric wear increases in approximately linear proportion to head diameter, about 4% to 6% per millimeter of head size increase. Other studies have demonstrated that bearing surface wear is far greater than backside wear[7] (at least in reasonably well-fixed liners), that thin polyethylene in combination with nonuniform backing support leads to increased wear, and that the type of cup fixation (cemented versus noncemented) has no direct kinetic influence on bearing surface wear behavior.[9] Small curvature radius mismatches, in the range of a few tenths of a millimeter between the acetabular liner and the femoral head, have no appreciable consequences in the long term; however, they do increase wear rates for the first few hundred thousand cycles as the polyethylene surface is reshaped to match the head's smaller radius of curvature.[10]

The vast improvement in abrasive and adhesive wear performance achieved with the new elevated cross-linked polyethylenes has the potential to change the entire landscape of metal-on-polyethylene hip replacement. This development may redirect some of the current interest in alternative bearing designs. Marked decreases in the intrinsic bearing couple wear coefficient make a relative wear penalty of 4% to 6% per millimeter of head size increase seem relatively innocuous, because absolute wear magnitude is so low. It now seems feasible to entertain head diameters well above the previous 32-mm consensus limit, in hopes of reducing dislocation by increasing range of motion prior to head-neck impingement. This of course implies a correspondingly thinner liner and presumably much higher polyethylene stresses for those impingement incidents that might still occur. Although very little information is available on the true incidence of impingement, one recent series cataloged grossly apparent evidence of rim denting on 66% of 180 retrieved liners.[11] How much the incidence of impingement—let alone frank dislocation—might actually be reduced by specific increases of head/neck range of motion is open to debate, especially since there is such great variability among individuals in dislocation-prone postural maneuvers.[12] That trade-off seems important, because elevated cross-linking carries a process-specific penalty in terms of reduced polyethylene ductility, toughness, and ultimate strength, a consideration which thus far has discouraged its application in the more fatigue-intense total knee replacement setting. How well the new elevated cross-linked polyethylenes will tolerate the occasional but truly enormous stress concentrations at impingement sites[13] remains to be seen. Their fatigue performance at sites of elevated stress near locking mechanisms also merits close monitoring.

Another set of design issues relates to unintentional rather than intentional influences on wear behavior. There has been striking intersubject variability in all total hip designs for which patient cohort wear rates have been measured. Activity level differences[14] among patients presumably are implicated, but there are large side-to-side differences in wear rates even in patients with identical bilateral constructs.[15] Gait style[16] and component orientation[17] likely also contribute, but these factors differ mildly in the vast majority of patients. A far more random clinical factor, one which under laboratory condition accounts for enormous variability in wear rate, is the presence of scratches on the femoral head. Individual scratches in the micron depth range can cause global wear rate to increase several fold,[18] and the macroscopically apparent burnished regions routinely seen on retrievals contain hundreds or thousands of such scratches.[19]

Parametric trials using an Archard-type computational model with local regions of femoral head roughening[20] have demonstrated that this phenomenon can account for order-of-magnitude differences in volumetric wear rate, and for wear directions that differ by tens of degrees, ranges of variation that are comparable to clinical experience.[21] Studies with roughened femoral balls in hip simulators have also shown consistent and substantial wear rate increases,[22,23] although such increases have been less spectacular for protocols with preintroduced scratches than for those with the presence of actual third-body debris,[24] possibly because a continuing supply of "fresh" scratches is needed to overcome the self-limiting effects of wear polishing and polyethylene impaction into existing scratch troughs. The limited amount of information currently available about the scratch-susceptibility of the elevated cross-linked polyethylenes is equivocal.[22,25] Although elevated cross-linked polyethylene appears to substantially outperform traditional polyethylene when articulated against a comparably scratched counterface, the cross-linked material's performance against a scratched counterface is substantially worse than traditional polyethylene's performance against a polished counterface.

The list of design variables potentially responsible for elevated third-body particulate burden is extensive. Even the ideally pristine situation represented in a cemented monobloc design has several third-body sources—cement and cement radio-opacifier particles, bone particles, metal particles from fretting of trochanteric reattachment wires, and metal particles from burnishing of loose stems. Noncemented fixation and modularity add more sources to the list: the femoral cement/metal interface, especially for matte and precoated stems; fretting debris from Morse taper head connections or from modular stem junctions; broken-off porous coating fragments, either late or during impaction at insertion; hydroxyapatite particles; metal fragments from the liner-backing interface of metal-backed acetabular components; and fretting debris from screw attachment of noncemented metal-backed cups. All of these design features have very legitimate advantages, and few if any in their present form are necessarily optimized in terms of minimizing third-body debris production. To the extent that patients at the high end of cohort wear rate distributions are the ones at greatest risk for osteolysis/loosening, that third-body debris is a major cause of accelerated (outlier) wear, that third-body rich design features continue to dominate the hip replacement marketplace, and that the new elevated cross-linked polyethylenes are vulnerable to accelerated wear from counterface scratches, it seems reasonable to expect that the problem of osteolysis and late aseptic loosening will remain a substantial complication of total hip arthroplasty.

Relevance

Because different hip joint implant designs have different wear rates, it is important to identify the factors responsible for these performance differentials. Both laboratory wear simulators and computational (finite element) models have proven helpful in that regard. Although new bearing surface materials appear to offer substantially lower intrinsic wear rates than metal on traditional polyethylene, use of these new material combinations frequently involves new mechanical design considerations, the impact of which must be understood prior to widespread clinical application. Even with reductions in cohort-averaged wear rates, a small fraction of patients still experience aberrantly rapid wear; the responsible factors for such behavior (eg, bearing surface damage from third-body debris) need to be elucidated.

Future Directions for Research

The spectacular wear rate reductions apparently available from the new elevated cross-linked polyethylenes provide a strong basis for optimism. However, to capitalize on this potential and to avoid the setbacks that have all too often plagued such promising technological improvements in the past, the research community must proactively confront an entirely new set of issues. From the mechanical design perspective, the constitutive and failure behavior of this new class of materials must be established. Sophisticated and insightful preclinical design studies using state-of-the-art finite element technology that is rigorously validated with experimental studies are also needed. As infrastructure to support such work, continued collection of clinical and physical data is necessary to make those models and experiments as realistic as possible. This seems particularly critical for new designs (eg, very large head sizes) that represent marked departures from traditional clinical practice. Techniques that more accurately measure in-service wear are required, as are better epidemiologic data and better long-term follow-up to identify the reasons why specific patients fare much more poorly than cohort averages. Finally, unless the new polyethylenes turn out to exhibit an immunity to accelerated wear from counterface roughness, a concerted effort is needed to devise mechanisms to obviate the potentially devastating effects of third-body debris.

References

1. Dowd JE, Sychterz CJ, Young AM, Engh CA: Characterization of long-term femoral-head penetration rates: Association with and prediction of osteolysis. *J Bone Joint Surg Am* 2000;82:1102-1107.

2. Harris WH: Osteolysis and particle disease in hip replacement. *Acta Orthop Scand* 1994;65:113-123.

3. Archard JF: Contact and rubbing of flat surfaces. *J Appl Phys* 1953;24:981-988.

4. McKellop HA, Clarke IC: Evolution and evaluation of materials screening machines and joint simulators in predicting in vivo wear phenomena, in Ducheyne P (ed): *Functional Behavior of Orthopaedic Biomaterials, Volume II: Applications.* Boca Raton, FL, CRC Press, 1984.

5. Bragdon CR, O'Connon DO, Weinberg EA, et al: The effect of cyclic rate on the wear of conventional and highly crosslinked UHMWPE acetabular components using the Boston AMTI hip simulator. *Trans Orthop Res Soc* 1999;24:831.

6. Clarke IC, Fujisawa A, Jung H: Influence of THR ball diameter on polyethylene wear rates. *Trans Soc Biomat* 1993;19:57.

7. Kurtz SM, Ochoa JA, Hovey CB, White CV: Frontside vs backside wear in an acetabular component with multiple screw holes. *Trans Orthop Res Soc* 1999;24:54.

8. Maxian TA, Brown TD, Pedersen DR, Callaghan JJ: A sliding-distance-coupled finite element formulation for polyethylene wear in total hip arthroplasty. *J Biomech* 1996;27:687-692.

9. Maxian TA, Brown TD, Pedersen DR, McKellop HA, Liu B, Callaghan JJ: Finite element analysis of acetabular wear: Validation and backing and fixation effects. *Clin Orthop* 1997;344:111-117.

10. Maxian TA, Brown TD, Pedersen DR, Callaghan JJ: Adaptive finite element modeling of long-term polyethylene wear in total hip arthroplasty. *J Orthop Res* 1996;14:668-675.

11. Hall RM, Siney P, Unsworth A, Wroblewski BM: Prevalence of impingement in explanted Charnley acetabular components. *J Orthop Sci* 1998;3:204-208.

12. Pedersen DR, Yack HJ, Wahi R, Callaghan JJ, Brown TD: Kinematics of maneuvers commonly responsible for hip dislocation. *Proceedings of the 24th Annual Meeting of the American Society of Biomechanics.* 2000:141-142.

13. Scifert CF, Brown TD, Pedersen DR, Heiner AD, callaghan JJ: Development and physical validation of a finite element model of total hip dislocation. *Comparative Methods in Biomechanical & Biomedical Engineering* 1999;2:139-147.

14. Schmalzried TP, Callaghan JJ: Wear in total hip replacements: Current concepts review. *J Bone Joint Surg Am* 1999;81:115-136.

15. Schmalzried TP, Dorey FJ, McClung CD, Scott DL: The contribution of wear mechanism(s) to variability in wear rates. *Trans Orthop Res Soc* 1999;24:287.

16. Bennett D, Orr JF, Baker R: The influence of shape and sliding distance of movement loci of the femoral head on the wear of the acetabular cup. *Proceedings of the 24th Annual Meeting of the American Society of Biomechanics.* 2000:99-100.

17. Yamaguchi M, Bauer TW, Hashimoto Y: A three-dimensional analysis of polyethylene wear vector in vivo. *Trans Orthop Res Soc* 1997;22:786.

18. Dowson D, Taheri S, Wallbridge NC: The role of counterface imperfections in the wear of polyethylene. *Wear* 1987;119:277-293.

19. Jasty M, Bragdon CR, Leek K, Hanson A, Harris WH: Surface damage to cobalt chrome femoral head prostheses. *J Bone Joint Surg Br* 1994;76:73-77.

20. Nieman JC, Brown TD, Pedersen DR, Callaghan JJ: The effects of localized regions of head roughening on variability of UHMWPE wear in THA. *Trans Orthop Res Soc* 1998;23:77.

21. Shaver SM, Brown TD, Hillis SL, Callaghan JJ: Digital edge-detection measurement of polyethylene wear after total hip arthroplasty. *J Bone Joint Surg Am* 1997;79:690-700.

22. McKellop HA, Shen F: Wear of surface gradient crosslinked UHMWPE cups against damaged femoral balls. *Trans Orthop Res Soc* 2000;25:25.

23. Wang A, Polineni VK, Stark C, Dumbleton JH: Effect of femoral head roughness on the wear of ultrahigh molecular weight polyethylene acetabular cups. *J Arthroplasty* 1998;13:615-620.

24. Davidson JA, Poggie RA, Mishra AK: Abrasive wear of ceramic, metal, and UHMW-PE bearing surfaces from third-body bone, PMMA bone cement, and titanium debris. *Biomed Mater Eng* 1994;4:312-329.

25. Burroughs BR, Blanchet TA: Effects of sliding directionality and countersurface finish on wear of irradiated UHMWPE. *Proceedings of the 6th World Biomaterials Conference*. May 2000.

What are the wear mechanisms and what controls them?

The primary materials used in bearing surfaces of total joint replacements include ultra-high molecular weight polyethylene, cobalt alloys, titanium alloys, stainless steel alloys, alumina, and zirconia. In general, the relationship between the properties of these materials and the in vivo wear performance of joint replacement components has been difficult to establish because so little is known about how these properties (and numerous potentially confounding patient, surgeon, and design-related factors) affect wear mechanisms.[1-4]

Modes of Wear

Wear occurs in four modes,[5-7] depending on location (Fig. 1). Mode 1 is the only wear mode associated with joint articulation; modes 2, 3, and 4 occur at other nonintentional articulations as a function of prosthesis materials, design, and implementation parameters.

Mode 1 is an articulation between intended bearing surfaces. Examples include the femoral head and the acetabular cup of a total hip replacement, and the femoral condyle and the tibial plateau of a total knee replacement. Examples of mode 2, an articulation between a primary bearing surface and

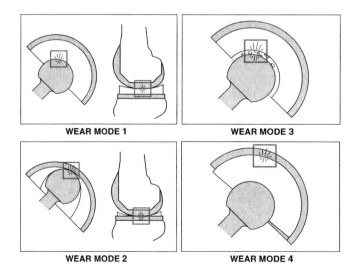

WEAR MODE 1   WEAR MODE 3

WEAR MODE 2   WEAR MODE 4

Figure 1 Modes of wear in orthopaedic joints.

a surface that was never intended to be a bearing surface, are the femoral head and the metal backing of an acetabular cup (eg, through a worn polyethylene acetabular liner) and the femoral condyle and the metal backing of a patellar component (eg, through a worn patellar button). Mode 3 is an articulation between intentional bearing surfaces in the presence of third-body components; examples include the femoral head and the acetabular cup in the presence of polymethylmethacrylate cement debris, metallic debris, hydroxyapatite particles, bone particles, or ceramic debris. One example of mode 4, an articulation between two nonbearing secondary surfaces, is backside wear caused by motion between the back of a polyethylene insert and the metallic tray of a total knee tibial component; another example is fretting wear between the trunion and cone of a modular femoral component of a total hip replacement.

Mechanisms of Wear

Wear can occur through five major mechanisms–adhesion, abrasion, third body, fatigue, and corrosion.[6-8] *Adhesive wear* occurs when the atomic forces occurring between the materials in two surfaces under relative load are stronger than the inherent material properties of either surface. For example, when there is relative motion between two surfaces, bonding of asperities occurs. Continued motion of the surfaces requires breaking the bond junctions. Each time a bond junction is broken, a wear particle is created, usually from the weaker material (Fig. 2). In orthopaedic joint replacements, adhesive wear usually occurs when small portions of the polyethylene surface adhere to the opposing metal bearing surface. The removal of polyethylene results in pits and voids so small that they may not be evident on visual inspection of the articulating surface.

The adhesive wear performance of both acetabular hip and tibial knee components has been related to the plastic flow behavior of polyethylene. In acetabular components, for example, the generation of submicron wear particles has been associated with local accumulation of plastic strain under multiaxial loading conditions until a critical or ultimate strain is reached.[9-11] Wear particles are released from the articulating surface following the accumulation of this critical plastic strain. Indeed, a plasticity-induced damage layer has been shown to develop at the articulating surface during hip simulator wear testing of both conventional and cross-linked polyethylene acetabular components.[12] The layer is associated with permanent reorientation of crystalline lamellae in the polyethylene morphology (Fig. 3).[13]

Abrasive wear occurs between surfaces of different relative hardness. In an abrasive wear mechanism, microroughened regions and small asperities on the harder surface locally plow through the softer surface (Fig. 4). Abrasive wear results in the softer material being removed from the track traced by the asperity during the motion of the harder surface.

Third-body wear is a form of abrasive wear that occurs when hard particles become embedded in a soft surface (Fig. 5). Examples of third bodies include metallic or bone particles embedded in a polyethylene bearing sur-

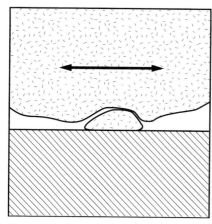

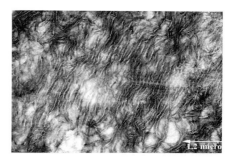

Figure 3 Plasticity-induced damage layer showing oriented lamellae. (Reproduced with permission from Kurtz SM, Rimnac CM, Pruitt L, Jewett CW, Goldberg V, Edidin AA: The relationship between the clinical performance and large deformation mechanical behavior of retrieved UHMWPE tibial inserts. Biomaterials 2000;21:283-291.)

Figure 2 Adhesive wear.

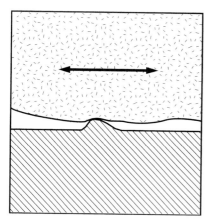

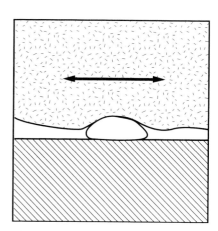

Figure 4 Abrasive wear.

Figure 5 Third-body wear.

face. The particle acts much like the asperity of a harder material in abrasive wear, removing material in its path. Hard third-body particles such as bone cement can produce damage to both the polyethylene articulating surface and the metallic alloy femoral bearing counterface.[14]

The extent of abrasive wear of polyethylene, metallics, and ceramics has been shown to be a function of the surface roughness of the metallic or ceramic counterface and the presence or absence of hard third-body particles.[14,15] In one in vitro hip simulator study, simulation of a roughened femoral head increased the amount of wear damage to the polyethylene, even in an elevated cross-linked polyethylene (although the overall wear rate was still dramatically lower than for conventional polyethylene articulating against a well-polished metallic counterface).[15] In other studies, isolated scratches more dramatically increased the wear rate than generalized roughness of the metallic counterface and could also change the wear performance ranking of various polyethylene formulations.[16] Thus the magnitude of the effect of surface roughness of the metallic counterface on overall wear rate remains controversial.

Fatigue wear occurs when surface and subsurface cyclic shear stresses or strains in the softer material of an articulation exceed the fatigue limit for that material. Because polyethylene is the weaker of the two materials in a bearing couple, fatigue wear damage to the polyethylene component dominates. Under these repeated or cyclic loading conditions, subsurface delamination and cracking can occur, eventually leading to the release of polyethylene particles (Fig. 6). Fatigue damage can range from small areas of pitting not apparent on visual inspection to macroscopic pits several millimeters in diameter to large areas of delamination that can encompass an entire tibial plateau.

Fatigue fracture mechanisms in tibial components have been directly related to the plastic flow parameters of polyethylene, such as yield stress and ultimate stress.[17] The performance of polyethylene components has also been associated with the presence of microscopic voids (so-called unconsolidated defects).[18-20] Thus, the plastic flow behavior and the presence of defects are believed to affect the clinical wear damage performance of polyethylene components. Implant retrieval analyses suggest that patient weight, activity level, and length of time of implantation are associated with the severity of surface damage of components.[21] Therefore, polyethylene fatigue fracture mechanisms have been suggested to contribute to certain forms (eg, pitting and delamination) of polyethylene surface damage.[22,23]

Damage modes such as accelerated fatigue wear, radial rim cracking, cup fracture, and delamination have been associated, at least in part, with oxidative degradation of the polyethylene.[2,3,24-27] In support of these reports, experimental studies demonstrated a significant decrease in fatigue and fracture resistance following oxidative degradation.[28,29] Delamination is probably not exclusively a consequence of subsurface oxidation, however. Research by Blunn and associates[30] supports the notion that damage to polyethylene tibial components is also dependent on joint kinematics. Delamination damage was observed on a flat polyethylene surface when a metal indenter had been sliding against it, but not for static loading or pure indenter rolling. Furthermore, the oxidative state and the quality of the polyethylene do not necessarily correlate with clinical wear performance. In a study of 92 retrieved Charnley acetabular components, no relationship was found between the radiographic wear rate measured while the components were implanted and either semiquantitative polyethylene measures (eg, the presence of a subsurface white band or the percentage area of unconsolidated particles) or changes in polyethylene density (an indirect measure of oxidative degradation).[4]

Corrosive wear is an indirect wear mechanism. A form of third-body wear, the liberated corrosive debris acts as an abrasive third body. Corrosive wear can also be considered an accelerating mechanism for corrosion itself, because the motion of an articulation can remove corrosive products and the

179

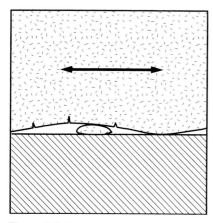

Figure 6 *Fatigue wear.*

protective passive layer sooner than interfaces with no relative motion. Liberation of corrosive products exposes a greater surface (with less protection against corrosion) to further corrosion, and hence accelerates the removal of even more material.

Adhesive, abrasive, and fatigue wear generally occur in both polyethylene acetabular hip and tibial knee components, although the relative contributions of each of these wear mechanisms differ in the two types of joints. Adhesive and abrasive wear generally dominate in polyethylene acetabular hip components, whereas fatigue wear is also an important wear mechanism in polyethylene tibial knee components. The forms of wear damage arising from the wear mechanisms described above include scratching, burnishing, abrasion, pitting, delamination, and embedded metallic or acrylic debris.[21]

Multiaxial Loading and Modeling Efforts

Efforts to understand the wear mechanisms and improve wear resistance of polyethylene have focused on the development of improved numerical models to predict the effect of load, geometry, and material properties on the stress and strain distributions occurring on and within joint replacement components. Early models incorporated loading only (monotonic behavior) without considering unloading, and were based on bilinear elastic or elastic-plastic approximations of the stress-strain behavior of polyethylene.[31-38] However, investigations of the unloading behavior and permanent plastic deformations in polyethylene have shown that classical plasticity theory greatly overpredicts the permanent strains on unloading. In fact, simulating cyclic loading in polyethylene components using conventional plasticity theory may lead to exaggerated predictions of residual strains and stresses.[39] Accordingly, modeling the cyclic loading behavior of polyethylene components may be more clinically relevant than modeling only monotonic behavior. The nonlinear behavior of polyethylene (eg, simultaneous recoverable and irrecoverable deformations on loading and unloading) also increases the importance of correctly simulating the kinematics of loading. Also, constitutive relationships for polyethylene should incorporate a continuous description of material response as the material transitions from linear viscoelastic to nonlinear viscoelastic to viscoplastic behavior.[40] The development of such constitutive models for conventional and cross-linked polyethylenes is currently an area of active research.[39,41,42]

Another important concentration of current research is on improved understanding of the more clinically relevant multiaxial large deformation behavior of polyethylene in predicting component performance, including surface damage. The small punch or miniaturized disk bend test has been proposed to characterize the equibiaxial tensile mechanical behavior of polyethylene using specimens measuring 0.5 mm in thickness.[43-45] Previously, indirect measures of the mechanical behavior of polyethylene, such as density, had been used to examine the relationship between mechanical performance and clinical wear performance of polyethylene.[34,46] A unique feature of the small punch test methodology is that it allows for the direct measurement of mechanical behavior from retrieved components that have dissimilar physical, chemical, and mechanical properties due to oxidative degradation, plasticity-induced alterations, or design-specified material property gradients. Results from these studies suggest that clinical wear performance may be predicted from the measured large deformation equibiaxial behavior of polyethylene.[45]

Relevance

Prevention of each of the wear mechanisms in each of the wear modes requires different materials and design considerations in the development of strategies for their avoidance. Design criteria should provide the best overall combination of materials, bearings, and surface finishes and treatments, derived from knowledge of existing strategies to protect against each of the wear mechanisms.

There is a notably better understanding today of the factors that influence adhesive, abrasive, and fatigue wear mechanisms in polyethylene joint components. However, the introduction of new polyethylene materials have still relied heavily on empirical in vitro hip simulator screening studies for acetabular hip components. With regard to polyethylene tibial knee replacement components, there is even less guidance on methods to predict the clinical performance of new polyethylene materials with respect to wear damage. Nevertheless, elevated cross-linked polyethylenes have been introduced into clinical practice not only for conventional acetabular hip components but also for more highly stressed, relatively thin (5-mm) acetabular hip components that articulate against a femoral head with a relatively large (38-mm) diameter. Cross-linked polyethylenes are also being considered for more highly stressed applications, such as those occurring in the less conforming articulating surface geometries in total knee replacements.[47] In all cases, these new materials (and designs) are being introduced into clinical practice without a fundamental understanding of the mechanical behavior of the material under the complex multiaxial loading conditions that these components undergo.

Future Directions for Research

Future research should include study of wear occurring at sites and in modes other than mode 1. Prevention of wear during intentional mode 1 articulation has received the most research attention, because wear is expected to occur at this site. However, unintentional mode 2, 3, and 4 articulations can also generate substantial wear. To fully address the issue of wear prevention, protection against wear mechanisms at unintended articulations must also be considered in future implant designs.

More accurate prediction of stress and strain distributions of polyethylene joint components using the finite element method is needed to ensure that joint replacement designs will benefit from the improved wear resistance of cross-linked polyethylenes (or other possible future material formulations) without sacrificing other important aspects of mechanical performance such as fatigue and fracture resistance. To this end, there is a need for a better understanding of the multiaxial mechanical behavior of conventional and cross-linked polyethylenes. In addition, constitutive relationships that capture the viscoelastic and viscoplastic characteristics of polyethylene need to be developed and implemented in time-dependent finite element analysis methodologies to improve prediction of the stress and strain distributions in polyethylene joint components.

Another important direction for future research is evaluation of the dynamic and time-dependent properties of implant materials and their articulations. Dynamic properties have received little attention, yet the materials are exposed to dynamic loading conditions. Material creep and relaxation during periods of relative inactivity (eg, during sleep) may influence accumulated stresses and could contribute to some of the discrepancies between in vitro wear simulations and in vivo results. More accurate predictions of the location and severity of permanent deformation to polyethylene components under in vivo loading conditions should help investigators develop better design paradigms to enhance long-term performance of joint replacements prior to their clinical introduction.

References

1. Wasielewski RC, Galante JO, Leighty RM, Natarajan RN, Rosenberg AG: Wear patterns on retrieved polyethylene tibial inserts and their relationship to technical considerations during total knee arthroplasty. *Clin Orthop* 1994;299:31-43.
2. Muratoglu OK, Mounib L, McGrory B, Bragdon CR, Harris JM, Harris WH: Anisotropic oxidation and radial cracks in retrieved acetabular components. *Trans Orthop Res Soc* 1998;23:307.
3. Walsh HA, Furman BD, Naab S, Li S: Determination of the role of oxidation in the clinical and in vitro fracture of acetabular cups. *Trans Soc Biomater* 1999;22:50.
4. Gomez-Barrena E, Li S, Furman BS, Masri BA, Wright TM, Salvati EA: Role of polyethylene oxidation and consolidation defects in cup performance. *Clin Orthop* 1998;352:105-117.
5. McKellop HA, Campbell P, Park PH, et al: The origin of submicron polyethylene wear debris in total hip arthroplasty. *Clin Orthop* 1995;311:3-20.
6. McKellop HA: Wear modes, mechanisms, damage, and debris: Separating cause from effect in the wear of total joint replacements, in Galante O, Rosenberg AG, Callaghan JJ (eds): *Total Hip Revision Surgery*. New York, NY, Raven Press, 1995, pp 21-39.
7. Schmalzreid TP, Callaghan JJ: Current concepts review: Wear in total hip and knee replacements. *J Bone Joint Surg Am* 1999;81:115-136.

8. Litsky AS, Spector M: Biomaterials, in Simon SR (ed): *Orthopaedic Basic Science*. Rosemont, IL, American Academy of Orthopaedic Surgeons, 1994, pp 447-486.
9. Wang A, Stark C, Dumbleton JH: Role of cyclic plastic deformation in the wear of UHMWPE acetabular cups. *J Biomed Mater Res* 1995;29:619-626.
10. Cooper JR, Dowson D, Fisher J: Macroscopic and microscopic wear mechanisms in ultra-high molecular weight polyethylene. *Wear* 1993;162-164:378-384.
11. Jasty M, Goetz DD, Bragdon CR, et al: Wear of polyethylene acetabular components in total hip arthroplasty: An analysis of one hundred and twenty-eight components retrieved at autopsy or revision operations. *J Bone Joint Surg Am* 1997;79:349-358.
12. Edidin AA, Pruitt L, Jewett CW, Crane D, Roberts D, Kurtz SM: Plasticity-induced damage layer is a precursor to wear in radiation-crosslinked UHMWPE acetabular components for total hip replacement. *J Arthroplasty* 1999;14:616-627.
13. Kurtz SM, Rimnac CM, Pruitt L, Jewett CW, Goldberg V, Edidin AA: The relationship between the clinical performance and large deformation mechanical behavior of retrieved UHMWPE tibial inserts. *Biomaterials* 2000;21:283-291.
14. Santavirta SS, Lappalainen R, Pekko P, Anttila A, Konttinen YT: The counterface, surface smoothness, tolerances, and coatings in total joint prostheses. *Clin Orthop* 1999;369:92-102.
15. McKellop H, Shen F-W, DiMaio W, Lancaster JG: Wear of gamma-crosslinked polyethylene acetabular cups against roughened femoral balls. *Clin Orthop* 1999;369:73-82.
16. Fisher J, Firkins P, Reeves EA, Hailey JL, Isaac GH: The influence of scratches to metallic counterfaces on the wear of ultra-high molecular weight polyethylene. *Proc Inst Mech Eng* 1995;209:263-264.
17. Pascaud RS, Evans WT, McCullagh PJ, FitzPatrick DP: Influence of gamma-irradiation sterilization and temperature on the fracture toughness of ultra-high-molecular-weight polyethylene. *Biomaterials* 1997;18:727-735.
18. Wrona M, Mayor MB, Collier JP, Jensen RE: The correlation between fusion defects and damage in tibial polyethylene bearings. *Clin Orthop* 1994;299:92-103.
19. Blunn GW, Lilley PA, Walker PS: Variability of the wear of ultra high molecular weight polyethylene in simulated TKR. *Trans Orthop Res Soc* 1994;19:177.
20. Walker PS, Blunn GW, Lilley PA: Wear testing of materials and surfaces for total knee replacement. *J Biomed Mater Res* 1996;33:159-175.
21. Hood RW, Wright TM, Burstein AH: Retrieval analysis of total knee prostheses: A method and its application to 48 total condylar prostheses. *J Biomed Mater Res* 1983;17:829-842.
22. Bartel DL, Rimnac CM, Wright TM: Evaluation and design of the articular surface, in Goldberg VM (ed): *Controversies of Total Knee Arthroplasty*. New York, NY, Raven Press, 1991, pp 61-73.
23. Elbert KE, Wright TM, Rimnac CM, et al: Fatigue crack propagation behavior of ultra high molecular weight polyethylene under mixed mode conditions. *J Biomed Mater Res* 1994;28:181-187.
24. Bohl JR, Bohl WR, Postak PD, Greenwald AS: The effects of shelf life on clinical outcome for gamma sterilized polyethylene tibial components. *Clin Orthop* 1999;367:28-38.
25. Sutula LC, Collier JP, Saum KA, et al: Impact of gamma sterilization on clinical performance of polyethylene in the hip. *Clin Orthop* 1995;319:28-40.
26. Won C-H, Rohatgi S, Kraay MJ, Goldberg VM, Rimnac CM: Effect of resin type and manufacturing method on wear of polyethylene tibial components. *Clin Orthop* 2000;376:161-171.

27. Bell CJ, Walker PS, Abeysundera MR, Simmons JM, King PM, Blunn GW: Effect of oxidation on delamination of ultrahigh-molecular-weight polyethylene tibial components. *J Arthroplasty* 1998;13:280-290.
28. Baldini TH, Rimnac CM, Wright TM: The effect of resin type and sterilization method on the static (J-integral) fracture resistance of UHMW polyethylene. *Trans Orthop Res Soc* 1997;22:780.
29. Baker DA, Hastings RS, Pruitt L: Compression and tension fatigue resistance of medical grade ultra high molecular weight polyethylene: the effect of morphology, sterilization,aging, and temperature. *Polymer* 2000;41:795-808.
30. Blunn GW, Walker PS, Joshi A, Hardinge K: The dominance of cyclic sliding in producing wear in total knee replacements. *Clin Orthop* 1991;273:253-260.
31. Bartel DL, Rawlinson JJ, Burstein AH, Ranawat CS, Flynn WF Jr: Stresses in polyethylene components of contemporary total knee replacements. *Clin Orthop* 1995;317:76-82.
32. Kurtz SM, Rimnac CM, Bartel DL: A predictive model for the tensile true stress-strain behavior of chemically and mechanically degraded ultra-high molecular weight polyethylene. *J Biomed Mat Res* 1998;43:241-248.
33. Kurtz SM, Rimnac CM, Santner TJ, Bartel DL: Exponential model for the tensile true stress-strain behavior of as-irradiated and oxidatively degraded ultra high molecular weight polyethylene. *J Orthop Res* 1996;14:755-761.
34. Kurtz SM, Bartel DL, Rimnac CM: Post-irradiation aging affects the stresses and strains in UHMWPE components for total joint replacement. *Clin Orthop* 1998;350:209-220.
35. Scifert CF, Brown TD, Pedersen DR, Callaghan JJ: Effects of acetabular component lip design on dislocation resistance in THA, in *Transactions of ASME Bioengineering (Winter)*. Dallas, TX, 1997.
36. Petrella AJ, Rubash HE, Miller MC: The effect of femoral component rotation on the stresses in an UHMWPE patellar prosthesis: A finite element study, in *Transactions of ASME Bioengineering (Winter)*. Dallas, TX, 1997.
37. Elbert K, Bartel D, Wright T: The effect of conformity on stresses in dome-shaped polyethylene patellar components. *Clin Orthop* 1995;317:71-75.
38. Estupinan JA, Bartel DL, Wright T: Residual stresses in ultra-high molecular weight polyethylene loaded cyclically by a rigid moving indenter in nonconforming geometeries. *J Orthop Res* 1998;16:80-88.
39. Estupiñán JA, Bartel DL, Wright TM: A micromechanically motivated constitutive model for the load and unload behavior of UHMWPE in total joint replacement. *Trans Orthop Res Soc* 2000;25:6.
40. Boyce MC, Chui C: Effects of heterogeneities and localization on polymer deformation and recovery, in Borst RD, Giessen EVD (eds): *Material Instabilities in Solids*. New York, NY, John Wiley & Sons, 1998.
41. Bergstrom J, Kurtz S, Rimnac C, Edidin A: Constitutive modeling of UHMWPE used in total joint replacements. *Trans Orthop Res Soc* 2001;26:1028.
42. Meyer R, Pruitt L: The effect of cyclic true strain on the morphology, structure, and relaxation behavior of ultra high molecular weight polyethylene. *Polymer* 2001;42:5293-5306.
43. Kurtz SM, Foulds JR, Jewett CW, Srivastav S, Edidin AA: Validation of a small punch testing technique to characterize the mechanical behavior of ultra-high molecular weight polyethylene. *Biomaterials* 1997;18:1659-1663.
44. Kurtz SM, Foulds JE, Jewett CW, Edidin AA: Small punch test for characterization of aged UHMWPE after gamma sterilization in air and nitrogen. *Trans Orthop Res Soc* 1998;23:361.

45. Kurtz SM, Rimnac CM, Pruitt L, Jewett CW, Goldberg V, Edidin AA: The relationship between the clinical performance and large deformation mechanical behavior of retrieved UHMWPE tibial inserts. *Biomaterials* 1999;21:283-291.

46. Bostrom MP, Bennett AP, Rimnac CM, Wright TM: The natural history of ultra high molecular weight polyethylene. *Clin Orthop* 1994;309:20-28.

47. Wang A, Polineni VK, Essner A, Sun DC, Stark C, Dumbleton JH: Effect of radiation dosage on the wear of stabilized UHMWPE evaluated by kip and knee joint simulators. *Trans Soc Biomater* 1997;20:394.

What material properties and manufacturing procedures influence wear mechanisms?

Wear phenomena in total joint replacements depend on a wide range of material properties. The biomechanical loading and load transfer over time are affected by the device material and design as well as functional aspects of the patient and the prosthesis. Thus the physical, mechanical, and chemical properties of a material and procedures involved in its manufacture that alter those properties all influence mechanisms of orthopaedic device wear.[1]

The orthopaedic literature contains a myriad of studies considering materials-related issues, including carbon and other reinforcements of polyethylene; heat pressing of polyethylene articulating surfaces; relative surface chemistry and roughness; ion-implantation of polyethylene and metallic bearing surfaces; extruding, compacting, machining, and direct and indirect molding of polyethylene; subsurface and bulk microstructures; molecular weight and distribution and phases; molecular chain length, orientation and alterations; fit of components specific to creep, fretting, and fatigue fracture; sterilization methods, residuals and oxidation; and, most recently, methods for elevated cross-linking and postirradiation processing.

Although it is quite well established that wear phenomena are influenced by material properties and manufacturing processes, multiple confounding conditions (eg, patient factors, surgical technique, material source, and processing) limit the identification of simple cause-effect relationships.[2]

There is considerable documentation of material modulus of elasticity (throughout the component, at or near the bearing surface, and within the bulk); compressive, tensile, and shear yield; tensile, creep, and fatigue strengths; fracture toughness; and wear failure modes.[3-5] Material alterations have affected performance. Several researchers have suggested that the alteration of elastic modulus by adding carbon fibers to polyethylene resulted in decreased localized biomechanical contact areas and increased stresses within the resulting polymer composite.[6] Property changes were then associated with wear phenomena of pitting and delamination. Others have proposed that the microstructural aspects of the composite (the orientations and distributions of carbon fibers), which lead to third-body particles and abrasive wear, have the most significant influence on in vivo performance.[7] Design choices can also affect wear behavior of an implant. Polyethylene component dimensions (eg, thickness of total knee tibial components) can cause unacceptable stress and strain distributions within the polyethylene subsurface zone.[3,4] The results of these finite element analyses led to recommendations of limits on thinness of components in order to avoid exceeding the material strength property limits.

Some industrial processing procedures, such as heat-pressing of polyethylene articulating surfaces, were shown to adversely influence component wear.[7] The localized alterations in mechanical and physical properties due to altered temperatures plus the changes in material flow and structural constitution were correlated with a surface zone in the material that was susceptible to biomechanical breakdown. Although the process was intended to enhance smoothness and intercomponent conformity, in vivo performance clearly demonstrated adverse wear performance. The altered modulus, microstructure, and strength within the surface zone may have resulted in the damage associated with in vivo functional loading. In response to these types of concerns, manufacturers have enhanced a number of processing steps and introduced quality control and assurance programs intended to minimize possible influences of unanticipated alterations in material characteristics.[7-10]

Laboratory studies of ion implantation, coatings, and environmental processing (primarily of titanium and cobalt alloy surfaces) demonstrated that metallic-polymeric interactions could be modified in an attempt to reduce wear.[11-12] Relative changes in surface roughness, hardness and in vivo wetting were shown to influence wear.[13-15] Changes in surface wetting and interfacial lubrication have also been proposed as explanations for better wear performance of ceramic and ceramic-like surfaces within total joint replacements.[16-18] In general, surface modifications decrease polyethylene wear, especially by minimizing third-body wear. Third-body particles remain an issue in metallic, ceramic, and polymeric interfaces within modular components, however, because debris can be generated from micromotion, fretting and corrosion.

The influence of polyethylene processing conditions (eg, extruding, machining, and molding) on material properties and wear performance remains complex and poorly understood. Analyses of explanted prostheses indicate that different parts of the devices have changed over time. Studies demonstrate a wide range of microstructural fusion defects (caused by inadequate compaction of the polyethylene powder when the material was fabricated), several types of extraneous particles, and the formation of residual stresses and strains (especially in extruded products) associated with processing polyethylene.[19-23] Debate continues on the property versus performance aspects of direct molded, premolded, and machined components. In vivo performance of total hip and knee devices has been affected by different manufacturing characteristics. Again, the industrial community has introduced a number of processing, sampling, and nondestructive testing protocols to minimize unanticipated structural alterations of articulating components.

Distributions and relative magnitudes of molecular weight, microstructural phases, and unanticipated features (eg, contaminant particles, fusion defects, and porosities) also have been associated with wear performance,[5,24] on the basis that alterations of the dense, uniform, standardized polyethylene product adversely influence wear. Once again, however, direct cause-effect correlations are complicated because multiple factors have been simultaneously altered in an attempt to improve performance. For example, one polyethylene product was modified in polymeric chain length, orientation, and distribution, resulting in somewhat altered material properties.[2] Subsequent studies indicated that the simultaneous introduction of microstructures and oxidation associated with gamma-radiation sterilization may have been more significant issues related to in vivo performance. Both material properties and processing factors affected in vivo wear performance. Some have suggested that the relative proportions of phases (eg, more crystalline regions and polymer chain orientations) are critical to both adhesive and abrasive wear characteristics.[1,25,26] Localized microscopic regions under the contact conditions of a total joint replacement could result in microfibular features extending from the articulating surfaces that in turn could fracture from the surface, leading to entry of microparticles into the contact area and causing abrasive wear.[1,18,27,28]

Biomechanical loading and the impact of localized stresses and strains exceeding the polymer creep and fatigue strengths of components have been studied in the context of polymer-metal component fit, component design, and manufacturing tolerance parameters. In general, change in shape due to creep and micromotion along backside interfaces have an adverse effect on in vivo wear behavior.[1,2,29]

A significant issue during the 1990s was the influence of sterilizing polyethylene components by gamma radiation in air followed by shelf aging in air before surgical implantation.[30-35] The formation of free radicals, oxidation, and altered material and performance properties have been associated with particle generation and subsequent inflammation and osteolysis. In this situation, material properties and industrial processing (sterilization) have a direct and strong influence on wear. Interestingly, some of the direct-molded polyethylenes introduced for joint replacement in decades past have shown limited breakdown due to wear, even though some oxidation apparently existed within the molded polyethylene.[5,8,9] Numerous studies have demonstrated the presence of environment-based transfers into polyethylene. Recently, complex esters from in vivo exposures were shown to influence material properties and wear phenomena by acting as a plasticizer, interacting as a barrier and absorber, and altering chemical/biochemical stabilities over time and use.[36] Possible influences of in vivo transfers specific to elevated cross-linked materials still need to be clarified relative to material properties as a function of the various manufacturing processes used for joint replacement products.

In response to concerns about polyethylene wear in hip and knee devices, over the past few years investigators have reintroduced old and introduced new elevated cross-linked polyethylenes in an effort to decrease in vivo wear.[36-44] Laboratory wear and joint simulator results indicate that polyethylene wear can be significantly reduced. At issue, however, are the material property alterations that depend upon the specific cross-linking (method and dose, for example) and processing (eg, heat-treating, melt-processing, sterilization, and packaging) procedures. Questions have been raised about the critical magnitudes of material strength properties and how these may influence overall clinical performance. Methods to eliminate or reduce oxidation of polyethylene subsurface regions have been a central focus of multiple investigations. Many manufacturers have made significant changes in the analysis, packaging, sterilization, and delivery of polyethylene joint replacement components. Assessing the impact of these changes will require ongoing clinical observation and evaluation.

Relevance

Basic and applied investigations have demonstrated that material properties and manufacturing processes do influence wear and mechanisms of wear. However, the impact of patient, surgical, device design, material, and manufacturing characteristics cannot be isolated; accordingly, simple cause-effect relationships are usually confounded and difficult to interpret. Unanticipated alterations in properties were often later shown to increase wear and decrease device longevity, emphasizing the need for careful consideration of ongoing material, design, and processing changes that are often intended to reduce in vivo wear. Most of the limitations that are now well recognized, however, were only determined after years of in vivo experience.

Future Directions for Research

Both material properties and manufacturing procedures directly and strongly influence wear. Alterations in physical, mechanical, and chemical material properties have been correlated with different magnitudes of polyethylene wear associated with adhesive, abrasive, and fatigue-type wear mechanisms. However, the overall literature presents a complex picture; most published investigations identify multiple confounding variables. The future direction for laboratory and clinical research and development should emphasize prospective protocols where applicable. In vitro, clinical in vivo, and device-retrieval data are needed to answer existing questions concerning wear and to define new issues and concerns as clinical experience is gained with new materials and designs intended to improve performance. The recommendations made at the recent NIH Technology Assessment Conference on implant retrieval[45] are indeed relevant to future research intended to establish effects of implant material and manufacturing processes on wear.

References

1. Wright TM, Goodman SB (eds): *Implant Wear: The Future of Total Joint Replacement*. Rosemont, IL, American Academy of Orthopaedic Surgeons, 1996, pp 98-102.

2. Kurtz S, Muratoglu O, Evans M, Edidin A: Advances in the processing, sterilization, and cross-linking of ultra-high molecular weight polyethylene for total joint arthroplasty. *Biomater* 1999;20:1659-1688.

3. Bartel DL, Bicknell VL, Wright TM: The effect of conformity, thickness, and material on stresses in ultra-high molecular weight components for total joint replacement. *J Bone Joint Surg Am* 1986;68:1041-1051.

4. Bartel DL, Rawlinson JJ, Burstein AH, et al: Stresses in polyethylene components of contemporary total knee replacements. *Clin Orthop* 1995;317:76-82.

5. Li S, Burstein AH: Ultra-high molecular weight polyethylene: The material and its use in total joint implants. *J Bone Joint Surg Am* 1994;76:1080-1090.

6. Wright TM, Astion DJ, Bansal M, et al: Failure of carbon fiber-reinforced polyethylene total knee-replacement components: A report of two cases. *J Bone Joint Surg Am* 1988;70:926-932.

7. *Transactions, Society for Biomaterials Symposium on Retrieval and Analysis of Surgical Implants and Biomaterials.* 1988;11:1-67.

8. Devanathan D, Bhambri D, Nazre A, Lin S: Characterization of compression molded ultra-high molecular weight polyethylene (UHMWPE). *Trans Soc Biomater* 1995;18:113.

9. Ramani K, Parasnis NC: Process-induced effects in compression molding of ultra-high molecular weight polyethylene (UHMWPE), in Gsell RA, Stein HL, Ploskonka JJ (eds): *Characterization and Properties of Ultra-High Molecular Weight Polyethylene.* West Conshohoken, PA, American Society for Testing and Materials, 1998.

10. Sauer WL, Anthony ME: Predicting the clinical wear performance of orthopaedic bearing surfaces, in Jacobs JJ, Craig TL (eds): *Alternative Bearing Surfaces in Total Joint Replacement.* West Conshohoken, PA, American Society for Testing and Materials, 1998.

11. Dearnaley G: Diamond-like carbon: A potential means of reducing wear in total joint replacements. *Clin Mater* 1993;12:237-244.

12. McKellop HA, Rostlund TV: The wear behavior of ion-implanted Ti-6A1-4V against UHMW polyethylene. *J Biomed Mater Res* 1990;24;1413-1425.

13. Lloyd AI, Noel RE: The effect of counterface surface roughness on the wear of UHMWPE in water and oil-in-water emulsion. *Tribol Int* 1988;21:83-88.

14. Like Q, Topoleski LDT, Parks NL: Surface roughness of retrieved CoCrMo alloy femoral components from PCA artificial total knee joints. *J Appl Biomat* 2000;53:111-118.

15. Peterson CD, Hillberry BM, Heck DA: Component wear of total knee prostheses using Ti-6Al-4V, titanium nitride coated Ti-6A1-4V, and cobalt-chromium-molybdenum femoral components. *J Biomed Mater Res* 1988;22:887-903.

16. Weightman BO, Simon SR, Paul IL, et al: Lubrication mechanisms of hip joint replacement prostheses. *J Lubric Technol* 1972;94:131-135.

17. McKellop H, Clarke I, Markolf K, et al: Friction and wear properties of polymer, metal and ceramic prosthetic joint materials evaluated on a multichannel screening device. *J Biomed Mater Res* 1981;15:619-653.

18. McKellop HA: Wear modes, mechanisms, damage, and debris: Separating cause from effect in the wear of total hip replacements, in Galante JO, Rosenberg AG, Callaghan JJ (eds): *Total Hip Revision Surgery.* New York, NY, Raven Press, 1995, pp 21-39.

19. Crugnola AM, Radin EL, Rose RM, et al: Ultrahigh molecular weight polyethylene as used in articular prostheses (a molecular weight distribution study). *J Appl Polym Sci* 1976;20:809-812.

20. Wrona M, Mayor MB, Collier JP, Jensen RE: The correlation between fusion defects and damage in tibial polyethylene bearings. *Clin Orthop* 1994;299:92-103.

21. Muratoglu OK, Jasty M, Harris WH: High resolution synchrotron infra-red microscopy of the structure of fusion defects in UHMWPE. *Trans Orthop Res Soc* 1997;22:773.

22. Li S, Chang JD, Barrena EG, et al: Nonconsolidated polyethylene particles and oxidation in Charnley acetabular cups. *Clin Orthop* 1995;319:54-63.

23. Wasielewski RC, Galante JO, Leighty RM, et al: Wear patterns on retrieved polyethylene tibial inserts and their relationship to technical considerations during total knee arthroplasty. *Clin Orthop* 1994;299:31-43.

24. Cooper JR, Dowson D, Fisher J: Macroscopic and microscopic wear mechanisms in ultra-high molecular weight polyethylene. *Wear* 1993;162:378-384.

25. Grood ES, Shastri R, Hopson CN: Analysis of retrieved implants: Crystallinity changes in ultrahigh molecular weight polyethylene. *J Biomed Mater Res* 1982;16:399-405.

26. Champion AR, Li S, Saum K, et al: The effect of crystallinity on the physical properties of UHMWPE. *Trans Orthop Res Soc* 1994;19:585.

27. McKellop HA, Campbell P, Park SH, et al: The origin of submicron polyethylene wear debris in total hip arthroplasty. *Clin Orthop* 1995;311:3-20.

28. Davidson JA, Poggie RA, Mishra AK: Abrasive wear of ceramic, metal, and UHMW-PE bearing surfaces from third-body bone, PMMA bone cement, and titanium debris. *Biomed Mater Eng* 1994;4:213-229.

29. Engh GA, Dwyer KA, Hanes CK: Polyethylene wear of metal-backed tibial components in total and unicompartmental knee prostheses. *J Bone Joint Surg Br* 1992;74:9-17.

30. Nusbaum HJ, Rose RM: The effects of radiation sterilization on the properties of ultrahigh molecular weight polyethylene. *J Biomed Mater Res* 1979;13:557-576.

31. Greer KW, Schmidt MB, Hamilton JV: The hip simulator wear of gamma-vacuum, gamma-air, and ethylene oxide sterilized UHMWPE following a severe oxidative challenge. *Trans Orthop Res Soc* 1998;23:52.

32. Collier JP, Sutula LC, Currier BH, et al: Overview of polyethylene as a bearing material: Comparison of sterilization methods. *Clin Orthop* 1997;333:76-86.

33. Rimnac CM, Klein RW, Betts F, et al: Post-irradiation aging of ultra-high molecular weight polyethylene. *J Bone Joint Surg Am* 1994;76:1052-1056.

34. Currier BH, Currier JH, Collier JP, Mayor MB, Scott RD: Shelf life and in vivo duration: Impacts on performance of tibial bearings. *Clin Orthop* 1997;342:111-122.

35. McKellop H, Shen FW, Ota T, Lu B, Wiser II, Yu E: The effect of sterilization method, calcium stearate and molecular weight on wear of UHMWPE acetabular cups. *Trans Soc Biomater* 1997;20:43.

36. Placko H: *In Vivo Aging of UHMW-PE in Total Joint Components.* Birmingham, AL, University of Alabama at Birmingham, 2000. PhD dissertation.

37. Streicher RM: Ionizing irradiation for sterilization and modification of high molecular weight polyethylenes. *Plast Rubber Process Appl* 1988;10:221-229.

38. Essner A, Polineni VK, Wang A, Stark C, Dumbleton JH: Effect of femoral head surface roughness and cross-linking on the wear of UHMWPE acetabular inserts. *Trans Soc Biomater* 1998;21:4.

39. Oonishi H, Takayama Y, Tsuji E: The low wear of cross-linked polyethylene socket in total hip prostheses, in Wise DL, Trantolo DJ, Altobelli DE, Yaszemski MJ, Gresser JD, Schwartz ER (eds): *Encyclopedic Handbook of Biomaterials and Bioengineering: Part A. Materials.* New York, NY, Marcel Dekker, 1995.

40. Wroblewski BM, Siney PD, Dowson D, Collins SN: Prospective clinical and joint simulator studies of a new total hip arthroplasty using alumina ceramic heads and cross-linked polyethylene cups. *J Bone Joint Surg Am* 1996;78:280-285.

41. Goldman M, Pruitt L: A comparison of the effects of gamma radiation and plasma sterilization on the molecular structure, fatigue resistance, and wear behavior of UHMWPE. *J Biomed Mater Res* 1998;40:378-384.

42. Jasty M, Bragdon CR, O'Connor DO, Muratoglu OK, Premnath V, Merrill E: Marked improvement in the wear resistance of a new form of UHMWPE in a physiologic hip simulator. *Trans Orthop Res Soc* 1997;43:785.

43. Bragdon CR, O'Connor DO, Muratoglu OK, et al: A new polyethylene with undetectable wear at 12 million cycles. *Trans Soc Biomater* 1998;21:2.

44. Muratoglu OK, Cook JL, Jasty M, Harris WH: A novel technique to measure the cross-link density of irradiated UHMWPE. *Trans Orthop Res Soc* 1998;23:782.

45. NIH Technology Assessment Conference: Improving Medical Implant Performance Through Retrieval Information: Challenges and Opportunities. NIH, January 10-12, 2000. (State of the Science Statement available online at http://odp.od.nih.gov/consensus/ta/019/019_statement.htm)

What modifications can be made to materials to improve wear behavior?

New Forms of Polyethylene

The load-bearing articulating surface materials used in total joint arthroplasty are metallic alloys, ceramics, and polymers. The articulating couples of primary concern—those that generate considerable amounts of wear leading to osteolysis—include ultra-high molecular weight polyethylene (eg, acetabular liners or tibial knee inserts). Accordingly, in the past decade most research has been focused on improving the wear resistance of polyethylene.

Consideration of appropriate modifications to polyethylene to improve its wear behavior requires an understanding of the wear mechanisms that occur in total hip and total knee replacements. In a series of 128 surgically and autopsy-retrieved acetabular liners, Jasty and associates found that the polyethylene surfaces underwent large-strain plastic deformation during in vivo use.[1,2] They hypothesized that numerous elongated fibrils indicative of large strain deformation (Fig. 1) lead to strain hardening of the material in the sliding direction, but weakening of the material in the transverse direc-

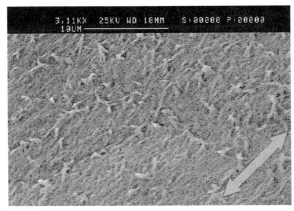

Figure 1 *Typical fine fibrils on the articulating surface of a retrieved acetabular liner. The fibrils indicate the presence of large-strain plastic deformation mainly in the flexion-extension direction. The arrow indicates the direction of flexion-extension. (Reproduced with permission from Muratoglu OK, Bragdon CR, O'Connor DO, Jasty M, Harris WH: A highly cross-linked UHMWPE: Expanded potential for total joint arthroplasty, in Rieker C, Oberholzer S, Wyss U (eds): World Tribology Forum in Arthroplasty. Bern, Switzerland, Hans Huber, 2001, pp 245-262.)*

tion. The oriented surface layer then breaks up during motions that are at an angle to these principal motion directions, liberating micron- and submicron-sized wear particles. These crossing motions are induced by the abduction/adduction and internal/external rotation of the hip. The implication is that the ability of the material to undergo large-strain plastic deformation at the surface in one direction and then to fracture as oriented fibrils under subsequent multidirectional motion producing cross shear strains is essential to polyethylene wear in hip replacements.[1,3,4]

The wear mechanisms of tibial knee inserts differ significantly from those of the acetabular liners. The major wear mechanisms in tibial knee inserts have been reported to be subsurface cracking, pitting, and delamination induced by the combination of peak subsurface stresses and oxidative embrittlement.[5-9] In order to increase delamination resistance in knee replacements, the oxidative stability of polyethylene must be improved. The abrasive and adhesive wear that dominate in hip replacements are also active in tibial knee inserts but to a lesser degree, leading to debris and periprosthetic osteolysis.[10] In knee replacements, the anterior-posterior translation combined with the flexion-extension motion induces the large-strain plastic deformation. The internal-external rotation then generates the required transverse motion leading to adhesive and abrasive wear (Fig. 2).[11,12]

One method for limiting large strain plastic deformation of the surface during articulation is cross-linking, the formation of covalent bonds between the polymeric chains; this inhibits chain mobility and reorientation, and hence wear. Cross-linking can be achieved by exposing polyethylene to ionizing radiation,[13] or to chemicals such as peroxides[14] or silanes.[15-17] The cross-links are formed by the reaction of the free radicals generated by these methods, leading to the creation of predominantly interchain covalent bonds.

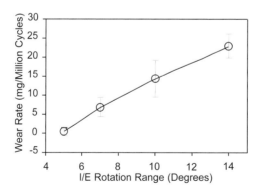

Figure 2 The wear rate of conventional tibial knee inserts as a function of internal/external (I/E) rotation on the AMTI knee simulator. Note that the I/E rotation has a strong effect on the multidirectional motion and hence wear. (Reproduced with permission from Muratoglu OK, Bragdon CR, O'Connor DO, Jasty M, Harris WH: A highly cross-linked UHMWPE: Expanded potential for total joint arthroplasty, in Rieker C, Oberholzer S, Wyss U (eds): World Tribology Forum in Arthroplasty. Bern, Switzerland, Hans Huber, 2001, pp 245-262.)

The use of polyethylenes cross-linked by silane chemistry was proposed to minimize polyethylene creep and hence the penetration of the femoral ball into the acetabular socket.[15-18] Today, however, cross-linking of polyethylene is pursued exclusively through exposure to ionizing radiation. Irradiation of polyethylene leads to stable cross-linked networks, provided that the residual free radicals generated during irradiation are eliminated in a postirradiation thermal treatment.[19,20]

With ionizing radiation, free radicals are formed through the radiolytic cleavage of C-H and C-C bonds in polyethylene. These free radicals recombine with each other and form crosslinks primarily in the amorphous portion of the polymer. The free radicals generated in the crystalline phase become trapped predominantly at the crystalline/amorphous interface and threaten the long-term oxidative stability of the material. The effects of residual free radicals on long-term embrittlement have been well documented for polyethylene that is gamma-sterilized and stored in air,[9,21] but is also of concern in the new methods intended to further elevate the cross-linking of polyethylene.[19,22]

Several methods have been developed for elevated radiation cross-linking of polyethylene that emphasize reduction or elimination of residual free radicals.[22] These methods can be grouped in three categories: irradiation at room temperature using gamma or electron beam irradiation followed by melting,[22,24] irradiation at an elevated temperature with adiabatic heating secondary to the electron beam irradiation followed by melting,[20] and irradiation at room temperature using gamma or electron beam irradiation followed by annealing below the melting transition.[25,26] After all three methods, the prostheses are machined, packaged, and sterilized. Sterilization options are ethylene oxide, gas plasma, or gamma irradiation; the first two methods do not generate additional residual free radicals in the polymer, while gamma sterilization leads to generation of residual free radicals.

Several factors have been shown to influence the properties of these new products, including the type of polyethylene resin used and how it was formed into the final geometry, the source of irradiation, the dose of irradiation, the environment in which the irradiation takes place, and postirradiation heat treatments to minimize oxidation. The current group of elevated cross-linked products are irradiated between 5 and 10 Mrads. The manufacturing processes of the commercial elevated cross-linked polyethylenes are summarized in Table 1.

The wear resistance depends primarily on the radiation dose level. Up to about 10 Mrads, the wear rate decreases with increasing radiation dose level.[22,24] The oxidative stability depends on the concentration of residual free radicals. The free radicals generated during the initial irradiation step are substantially eliminated by subsequent thermal treatment. The elimination of free radicals is not as effective when annealing is carried out below the melting temperature of the polymer.[23] In addition, if an implant made of elevated cross-linked and thermally annealed polyethylene is subsequently sterilized with gamma irradiation, residual free radicals would be generated, adversely affecting oxidative stability.

Table 1 Comparison of new cross-linked thermally stabilized polyethylenes

Name and Manufacturer	FDA Approved? (June 2000)	Radiation Type and Dose	Thermal Stabilization
Marathon™ DePuy, Inc	Yes	Gamma radiation to 5 Mrads at room temperature	Remelted at 155° C for 24 hours
XLPE™ Smith & Nephew-Richards, Inc	Yes	Gamma radiation to 10 Mrads at room temperature	Remelted at 150° C for 2 hours
Longevity™ Zimmer, Inc	Yes	Electron beam radiation to 10 Mrads at 40° C	Remelted at 150° C for about 6 hours
Durasul™ Sulzer, Inc	Yes	Electron beam radiation to 9.5 Mrads at 125° C	Remelted at 150° C for about 2 hours
Crossfire™ Stryker-Osteonics-Howmedica, Inc	Yes	Gamma radiation to 7.5 Mrads at room temperature	Annealed at about 120° C for a proprietary duration
Aeonian™ Kyocera, Inc	Approved in Japan, not yet in USA	Gamma to 3.5 Mrads at room temperature	Annealed at 110° C for 10 hours

Adapted with permission from McKellop HA: Bearing surfaces in total hip replacements: State of the art and future developments. Instr Course Lect 2001;50:165-179.

Although effective in reducing hip simulator wear rates, increasing the cross-linking in polyethylene can also change other physical properties. In a parametric study of 34 different cross-linking conditions, including variations in source of irradiation (gamma or electron beam), dose (0 to 50 Mrads), and postirradiation melting, yield strength could be reduced up to

Table 1 continued

Final Sterilization	Manufacturer's Rationale
Gas plasma	5 Mrads gamma cross-linking provides about 85% wear reduction (theoretically below the threshold for lysis) while preserving other mechanical properties. Remelting substantially eliminates free radicals and substantially prevents oxidative degradation.
Ethylene oxide	10 Mrads gamma provides wear reduction below their detection limit. Remelting substantially eliminates free radicals and substantially prevents oxidative degradation.
Gas plasma	10 Mrads provides about 89% wear reduction with femoral head sizes of 22 to 40 mm. The increased temperature and dose rate of irradiation preserve the mechanical properties. Remelting substantially eliminates free radicals and substantially prevents oxidative degradation.
Ethylene oxide	9.5 Mrads provides wear below their detection limit with femoral head sizes of 22 to 46 mm. The increased temperature and irradiation dose rate preserve the mechanical properties. Remelting substantially eliminates free radicals and substantially prevents oxidative degradation.
Gamma at 2.5 to 3.5 Mrads while packaged in nitrogen	Depending on actual sterilization dose, total gamma cross-linking dose may vary from 10 to 11 Mrads, providing about 90% wear reduction. Annealing provides a different balance of material properties than remelting. Resultant material has substantial free radicals, but oxidation is limited by sterilization and storage in nitrogen.
Gamma at 2.5 to 4 Mrads while packaged in nitrogen	Depending on actual sterilization dose, total gamma cross-linking dose may vary from 6 to 7.5 Mrads. Wear data not yet available. Annealing provides a different balance of material properties than remelting. Resultant material has substantial free radicals, but oxidation is limited by sterilization and storage in nitrogen.

30%, elongation to break by 45%, and modulus lowered up to 27% (depending on the combination of cross-linking factors) compared to ram-extruded 4150HP gamma-sterilized in air at 2.5 Mrads. Although these properties are changed in the elevated cross-linked products, the values for the physical properties are still in line with ASTM guidelines for ultra-high molecular weight polyethylene. However, these guidelines are not related to device performance but are intended only as descriptors of the material. No two commercially available elevated cross-linked products are produced in the same manner. Although it is not currently possible to directly associate changes or differences in material properties to clinical performance, these varying manufacturing conditions may provide different clinical results.

Additionally, the fracture toughness of polyethylene, as assessed by J integral measurements, was adversely affected by different cross-linking methods (Fig. 3).[27] As the irradiation dose increased, the J integral fracture toughness decreased. This same trend of loss of fracture toughness was found in electron beam irradiated samples. The loss of fracture toughness due to increased cross-linking is of some concern. Although fracture in acetabular cups is generally rare, there are a number of reports of acetabular

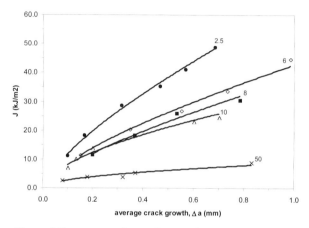

Figure 3 Fracture toughness, J integral versus iradiation dose.

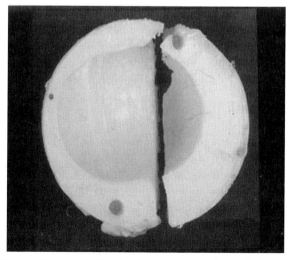

Figure 4 Fractured all polyethylene, cemented acetabular cup.

cup fracture in the literature. These reports include both all-polyethylene cemented cups (Fig. 4) and modular metal-backed cups. If there is currently some, albeit small, incidence of fracture with traditional forms of polyethylene that have been cross-linked from sterilization by gamma irradiation between 2.5 and 4 Mrads, reduction of fracture toughness of 40% or more (Fig. 3) may lead to an increase in the number of fracture-related failures. (Of course, oxidative degradation may have contributed to the fractures reported in conventional polyethylene cups.) This concern may be further accentuated by design issues, especially in metal-backed cups in which small polyethylene structures (Fig. 5) are used to hold or lock the polyethylene liner in place or in cases where the polyethylene cup is not fully supported, leading to rim loading (Fig. 6). Additionally, the response of an elevated cross-

Figure 5 *Thin polyethylene locking tab.*

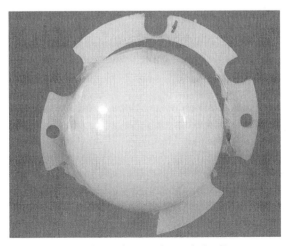

Figure 6 *Rim-fractured acetabular liner.*

linked polyethylene acetabular cup with reduced fracture toughness to femoral neck impingement is unknown. Retrieved acetabular components made from conventional polyethylene often show fracture of the polyethylene secondary to impingement (Fig. 7).[28]

Only a few clinical reports have been published on these new materials. Short-term (1- to 2-year) follow-ups showed that clinical wear rates are the same as for metal on conventionally (2.5 to 4 Mrads) irradiated polyethylene. This result should be expected because the creep properties are unchanged by cross-linking and the actual amount of wear occurring in this time period may be very small.[29]

Molding of Polyethylene in Total Knee Replacements

The Insall-Burstein (IB), Miller-Galante (MG), and AGC total knee systems have all demonstrated excellent long-term tibial insert survival despite their significant variances in contact stresses caused by differences in bearing surface geometries and other design features. In studies of retrieved components, these three designs showed no delamination or fracture.[30-33] One key similarity of these knee systems is that the tibial inserts were directly molded from 1900 UHMWPE resin. From a materials standpoint, the direct molding process imparts characteristics to polyethylene that are not achieved by either ram extrusion or compression molding of large sheets. Some components made by direct molding do not show significant oxidation-induced embrittlement 10 years after gamma irradiation in air. This is in stark contrast to the behavior of components made by other manufacturing methods and other resins that demonstrate continuous oxidative degradation beginning immediately after irradiation.[34] The J integral fracture toughness of directly molded polyethylene is also higher than that of samples made by extrusion or compression molding.[27]

The similar wear performance of the IB, MG, and AGC knee systems suggests that directly molded polyethylene may be more useful in higher stress applications than polyethylene fabricated by other methods. Thus,

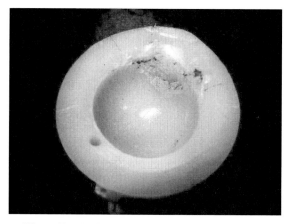

Figure 7 *Polyethylene fracture due to impingement.*

directly molded polyethylene may benefit any device or design that experiences high stresses, such as unicondylar knees and extremity joints.

The benefits of elevated cross-linked polyethylene in total knee replacements are unclear. The loss of fracture toughness with higher levels of cross-linking is critical because of the high stresses experienced by tibial and patellar contact surfaces. Fatigue-related damage phenomena such as pitting, delamination, and fracture associated with initiation and propagation of cracks are much more prevalent in total knee replacements than in total hip replacements.

Other Design Improvements

Design improvements in total knee replacements have focused on reducing the stresses in the polyethylene. However, the long-term clinical success of directly molded implants with relatively nonconforming bearing surfaces (ie, the MG and AGC knee systems) suggests that this traditional means of evaluating performance may not be an effective predictor of clinical failure and that designs with higher contact stress may perform satisfactorily. Thus design options may be broader with the use of directly molded resins, allowing for increased flexion angles, more natural kinematics, and thinner components than might be considered based on stress predictions alone.

Backside wear of tibial inserts against the metal tray may be a significant source of polyethylene wear in total knee replacements and in some cases may actually be the largest source of debris.[35-37] Backside wear debris may be minimized by using one of the new forms of elevated cross-linked polyethylene in conjunction with other design changes.

Ceramic-on-Ceramic Total Hip Replacements

Ceramic-on-ceramic total hips have been used since the late 1970s. Early clinical results showed higher failure rates than metal-on-polyethylene systems. The major indications for revision were ceramic fracture and aseptic loosening.

Contemporary ceramic-on-ceramic devices differ from these earlier devices in both design and materials. Physical and fracture properties of alumina ceramics have been improved through better control of grain size. As the grain size has fallen, the fracture toughness of the ceramic has improved. Today the incidence of ceramic fracture is rare, estimated to be on the same order as fracture of hip stems. Ceramic acetabular components are now modular, with a metal shell and ceramic insert. Although the current ceramics are more fracture-resistant, their fracture toughness is dramatically lowered if a chip or notch is introduced in the component, such as might be caused by impingement. Nonetheless, a ceramic-on-ceramic total hip system has recently received FDA approval.

Metal-on-Metal Total Hips

The cobalt alloy metal-on-metal articulating couples provide another example of a system that generates essentially zero wear in laboratory simulations. Current devices pair metal acetabular shells with an insert made of polished metal. Ongoing efforts to improve this system include surface modifications, such as ion and electron beam irradiation and altered surface chemistries, and hardening and polishing the metals with thermomechanical processes. Efforts are also underway to significantly change surface properties by the application of diamond-like and diamond coatings.

Relevance

The marked reduction in wear of polyethylene following radiation cross-linking addresses the most important problem limiting the longevity of total hip replacements and an important issue in total knee replacement. The new elevated cross-linked materials have significantly improved wear resistance and exhibit greatly enhanced resistance to oxidation. The reduction in wear has strong potential to reduce the clinical rate of osteolysis, which should lower the incidence of revision hip replacement surgery, even in young and active patients. The increase in oxidative resistance has the potential to eliminate delamination in tibial knee inserts. Increased wear resistance offers the possibility of using larger femoral head sizes, thus increasing stability, enhancing range of motion, and reducing the rate of dislocation. Thinner polyethylene liners (as low as 4 mm) may also be possible, satisfying the need for large head sizes in small implants (eg, acetabular components with 28-mm inner diameters and 37-mm outer diameters), providing improved stability and range of motion in patients with juvenile rheumatoid arthritis and developmental dysplasia of the hip and in patients with smaller stature. The catastrophic failure of polyethylene patellar buttons secondary to oxidation-induced embrittlement could also be significantly reduced with the use of these new materials.

Future Directions for Research

Prospective, randomized double-blind clinical studies using improved in vivo wear detection techniques such as radiostereometric analysis are needed to determine the efficacy of elevated cross-linked polyethylenes in reducing wear and osteolysis. Additional larger scale clinical follow-up studies using two- or three-dimensional radiographic wear measurements and observations on retrieved implants should be conducted to further evaluate the performance of these new materials.

An obvious developmental effort is needed to improve the performance of joints other than hip and knee. The shoulder, elbow, and ankle markets are growing rapidly, yet an understanding of the performance of these joints lags that of hips and knees. These applications present a challenge to conventional forms of polyethylene because of the high loads applied over small contact areas. Implant fixation is more difficult given the anatomic constraints in small bones such as the glenoid and talus. These devices have achieved limited clinical success. Use of new forms of UHMWPE and development of new design concepts may improve the lifetime of these devices.

The future developments in ceramics are directed toward the continued improvement of properties. The two most recent activities include new forms of zirconia and alumina/zirconia ceramic composites.

As the level of polyethylene debris generation is reduced, attention needs to be focused on the next weakest link in the bone-implant construct. For cemented systems, the weak link may be the cement itself. Improvements in fatigue and fracture resistance may help to limit mechanical failures and subsequent loosening. The recent reclassification of bone cement by the FDA from a class III to a class II device lowers the regulatory requirements to commercial development of improved cement or cement systems.

References

1. Jasty MJ, Goetz DD, Lee KR, Hanson AE, Elder JR, Harris WH: Wear of polyethylene acetabular components in total hip arthroplasty: An analysis of 128 components retrieved at autopsy or revision operation. *J Bone Joint Surg Am* 1997;79:349-358.

2. Jasty M, Bragdon C, Jiranek W, Chandler H, Maloney W, Harris WH: Etiology of osteolysis around porous-coated cementless total hip arthroplasties. *Clin Orthop* 1994;308:111-126.

3. Wang A, Sun DC, Stark C, Dumbleton JH: Wear mechanisms of UHMWPE in total joint replacements. *Wear* 1995;181-183:241-249.

4. Kurtz SM, Pruitt LA, Jewett CW, Foulds JR, Edidin AA: Radiation and chemical crosslinking promote strain hardening behavior and molecular alignment in ultra high molecular weight polyethylene during multi-axial loading conditions. *Biomaterials* 1999;20:1449-1462.

5. Collier JP, Bauer TW, Bloebaum RD, et al: Results of implant retrieval from post-mortem specimens in patients with well-functioning, long-term total hip replacement. *Clin Orthop* 1992;274:97-112.

6. Cornwall GB, Bryant JT, Hansson CM, Rudan J, Kennedy LA, Cooke TD: A quantitative technique for reporting surface degradation patterns of UHMWPE components of retrieved total knee replacements. *J Appl Biomat* 1995;6:9-18.

7. Cameron HU: Tibial component wear in total knee replacement. *Clin Orthop* 1994;309:29-32.

8. Collier JP, Sperling DK, Currier JH, Sutula LC, Saum KA, Mayor MB: Impact of gamma sterilization on clinical performance of polyethylene in the knee. *J Arthroplasty* 1996;11:377-389.

9. Collier JP, Mayor MB, McNamara JL, Surprenant VA, Jensen RE: Analysis of the failure of 122 polyethylene inserts from uncemented tibial knee components. *Clin Orthop* 1991;273:232-242.

10. Peters P, Engh G, Dwyer K, Vinh T: Osteolysis after total knee arthroplasty without cement. *J Bone Joint Surg Am* 1992;74:864-876.

11. Muratoglu OK, Bragdon CR, O'Connor DO, Jasty M, Harris WH: A highly crosslinked UHMWPE: Expanded potential for total joint arthroplasty, in Rieker C, Oberholzer S, Wyss U (eds): *World Tribology Forum in Arthroplasty*. Bern, Switzerland, Hans Huber, 2001, pp 245-262.

12. Wang A, Sun DC, Stark C, Dumbleton IH: Effects of sterilization methods on the wear of UHMWPE acetabular components, in *Proceedings of the Fifth World Biomaterials Congress*. 1996, p 196.

13. Charlesby A: Cross-linking of polythene by pile radiation. *Proc Roy Soc Lond* 1952;A215:187-215.

14. de Boer J, Pennings A: Crosslinking of ultra-high molecular weight polyethylene in the melt by means of 2.5-dimethyl-2.5-bis(tert-butyldioxy)-3-hexyne: 2. Crystallization behaviour and mechanical properties. *Polymer* 1982;23:1944-1952.

15. Atkinson JR, Dowling JM, Cicek RZ, Materials for internal prostheses: the present position and possible future developments. *Biomaterials* 1980;1:89-96.

16. Atkinson JR, Cicek RZ: Silane crosslinked polyethylene for prosthetic applications: Part I. Certain physical and mechanical properties related to the nature of the material. *Biomaterials* 1983;4:267-275.

17. Atkinson JR, Cicek RZ: Silane crosslinked polyethylene for prosthetic applications: Part II. Creep and wear behavior and a preliminary molding test. *Biomaterials* 1984;5:326-335.

18. Wroblewski B, Siney P, Fleming P: Low-friction arthroplasty of the hip using alumina ceramic and cross-linked polyethylene: A ten-year follow-up report. *J Bone Joint Surg Br* 1999;81:54-55.

19. Muratoglu OK, Bragdon CR, O'Connor DO, Merrill EW, Jasty EM, Harris WH: Electron beam crosslinking of UHMWPE at room temperature, a candidate bearing material for total joint arthroplasty. *Trans Soc Biomater* 1997;20:74.

20. Muratoglu O, Bragdon C, O'Connor D, Jasty M, Harris W: A novel method of crosslinking UHMWPE to improve wear, reduce oxidation, and retain mechanical properties. *J Arthroplasty* 2001;16:149-160.

21. Sutula LC, Collier JP, Saum KA, et al: Impact of gamma sterilization on clinical performance of polyethylene in the hip. *Clin Orthop* 1995;319:28-40.

22. McKellop H, Shen FW, Lu B, Campbell P, Salovey R: Development of an extremely wear resistant ultra-high molecular weight polyethylene for total hip replacements. *J Orthop Res* 1999;17:157-167.

23. Muratoglu OK, Bragdon CR, O'Connor DO, et al: Comparison of wear behavior of four different types of crosslinked acetabular components. *Trans Orthop Res Soc* 2000;25:566.

24. Muratoglu OK, Bragdon CR, O'Connor DO, et al: Unified wear model for highly crosslinked ultra-high molecular weight polyethylenes (UHMWPE). *Biomaterials* 1999;20:1463-1470.

25. Taylor SK, Serekian P, Bruchalski P, Manley M: The performance of irradiation-crosslinked UHMWPE cups under abrasive conditions throughout hip joint simulation wear testing. *Trans Orthop Res Soc* 1999;24:252.

26. Manley MT: Crossfire polyethylene: A bearing material for reducing wear in total hip arthroplasty. Presented in Harvard University Hip Course, 1999, Cambridge, MA.

27. Duus LC, Walsh HA, Gillis AM, Noisiez E, Li S: The effect of resin grade, manufacturing method and cross linking on the fracture toughness of commercially available UHMWPE. *Trans Orthop Res Soc* 2000;25:544.

28. Shon WY, Wright TM, Baldini T, Peterson M, Salvati E: Impingement in total hip arthroplasty: A study of retrieved acetabular components. *Trans Orthop Res Soc* 2001;26:1070.

29. Furman B, Saunders M, Potier E, Li S: Effect of sterilization method and dose on the creep properties of cross linked UHMWPE, in *Transactions of the Sixth World Biomaterials Congress*. 2000, p 421.

30. Furman B, Ritter MA, Perone JB, Furman GF, Li S: Effect of resin type and manufacturing method on UHMWPE oxidation and quality at long aging and implant times. *Trans Orthop Res Soc* 1997;22:92.

31. Furman B, Awad JN, Chastain KE, Li S: Material and performance differences between retrieved machined and molded Insall Burstein type total knee prosthesis. *Trans Orthop Res Soc* 1997;22:643.

32. Scuderi GR, Insall JN, Windsor RE, Moran MC: Survivorship of cemented knee replacements. *J Bone Joint Surg Br* 1989;71:798-803.

33. Ritter MA, Worland R, Saliski J, et al: Flat on flat nonconstrained compression molded polyethylene total knee replacement. *Clin Orthop* 1995;321:79-85.

34. Furman BD, Li S: Effect of resin type on the oxidation of UHMWPE, in *Proceedings of the Fifth World Biomaterials Congress.* 1996, p 188.

35. Furman BD, Schmieg JJ, Bhattacharya S, Li S: Assessment of backside polyethylene wear in 3 different metal backed total knee designs. *Trans Othop Res Soc* 1999;24:149.

36. Griffin FM, Scuderi GR, Gillis AM, Li S, Jimenez E, Smith T: Osteolysis associated with cemented total knee arthroplasty. *J Arthroplasty* 1988;13:592-598.

37. Parks NL, Engh GA, Topoleski LD, Emperado J: Modular tibial insert micromotion: A concern with contemporary knee implants. *Clin Orthop* 1998;356:10-15.

What evidence is there for using alternative bearing materials?

The vast majority of hip prostheses have cobalt-chrome alloy femoral balls bearing against polyethylene acetabular cups that were sterilized with 2.5 to 4 Mrads of gamma radiation in air and stored in air, sometimes for several years. The wear rate and other mechanical properties of this combination now form the clinical baseline against which potentially improved bearing combinations are evaluated. The average clinical wear rate of polyethylene against cobalt-chrome alloy is in the range of 0.1 to 0.2 mm per year. However, this average includes implants that have accelerated wear rates due to excessive third-body damage to the bearing surfaces, radiation-induced oxidative degradation of the polyethylene, and other causes.[1,2] Thus, the inherent wear rate of a recently gamma-air sterilized polyethylene cup with a cobalt-chrome alloy ball, running under clean conditions, would be expected to be substantially lower than the clinical average wear rate, possibly as low as 0.05 mm per year. These factors must be taken into account when the wear performance of a new combination of bearing materials is evaluated in the laboratory in a joint simulator.

Potential improvements in hip bearing materials include either replacing polyethylene altogether, as with metal-on-metal and ceramic-on-ceramic bearings, or modifying the method of producing polyethylene components in order to minimize the effects of oxidative degradation, while inducing sufficient cross-linking to minimize wear.

Metal-Metal Bearings

Second-generation metal-metal implants include conventional total hips and surface replacements.[3-5] The first to be widely used clinically is the Metasul™ hip,[6,7] which recently received FDA approval for use in the United States. Hip simulator studies[8-10] and clinical retrievals[7,11,12] of modern metal-metal bearings have typically shown steady-state wear rates on the order of a few microns per million cycles. Metal-metal bearings can also self heal, as isolated surface scratches caused by third-body particles or subluxation damage are polished out during subsequent use.[13] The clinical performance of second-generation metal-metal hips has been comparable to that of

This chapter was adapted with permission from McKellop HA: Bearing surfaces in total hip replacements: State of the art and future developments. Instr Course Lect 2001;50:165-179.

conventional metal-polyethylene hips, and somewhat better than first-generation metal-metal hips.[7,14]

Clinical and laboratory wear studies have indicated that metal-metal bearings often exhibit 10 to 20 times greater wear-in during the initial 1 to 2 years of clinical use (Fig. 1) or the first 1 to 2 million cycles in a hip simulator.[7-9,11] In addition, some metal-metal bearings have exhibited extensive surface micropitting, possibly due to a fatigue-corrosion mechanism associated with the smaller carbides.[11,13] Although the high wear-in rate is a transitory phenomenon and the presence of the micropits does not appear to be associated with a high wear rate,[11] these are areas of potential improvement in metal-metal implants.

Ceramic Bearings

Ceramic-Polyethylene Bearings

Alumina and zirconia femoral balls have been widely used as bearing surfaces against polyethylene cups, and the majority of clinical studies have shown substantially lower polyethylene wear rates than with metal balls, with wear ratios ranging from 0.25 to 0.75.[2,15,16] However, a recent radiographic study reported little difference in polyethylene wear with alumina versus metal balls.[17] Another study reported unacceptably high rates of polyethylene wear, lysis, and loosening with an early zirconia ball design.[18] The differences in clinical results may be due in part to the influence of third-body particles. The greater hardness of ceramic balls renders them more resistant than metal balls to scratching by entrapped abrasive contaminants that can accelerate wear of the opposing polyethylene cup. The differences in wear performance might, therefore, be more reflective of differences

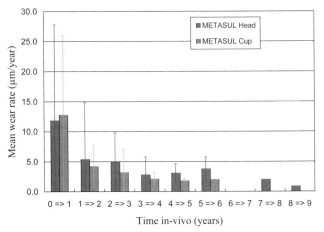

Figure 1 *Wear rate for retrieved second-generation metal-metal hip prostheses. (Reproduced with permission from Rieker CB, Köttig P, Schön R, Windler M, Wyss UP: Clinical tribological performance of 144 metal-on-metal hip articulations, in Rieker C, Windler M, Wyss U (eds): Metasul: A Metal-on-Metal Bearing. Bern, Switzerland, Hans Huber, 1999, pp 83-91.)*

in the amount of third-body contamination. Regardless of the bearing material used, care must be taken to minimize the formation of abrasive contaminants in vivo, for example by avoiding those porous coatings that tend to shed particles.

Alumina-Alumina Bearings

Like metal-metal bearings, the earliest designs of alumina-alumina hip bearings often had unacceptably high wear rates.[15,19-21] The causes included poor quality ceramic, edge contact of the cup on the ball due to inadequate range of motion and/or vertical cup placement, and other design shortcomings. Fortunately, over the past two decades substantial improvement has been made in prosthesis design, implantation technique, and the quality of the alumina[15,22] (such as higher purity, finer grain structure, and improved sintering techniques). Modern alumina-alumina bearings are not, however, immune to high wear.[23,24] In one recent series, severe wear and osteolysis were reported in 22% of the patients after an average follow-up of only 7.7 years.[25]

Gross fracture of a ceramic bearing component is a catastrophic failure, and the surgeon performing the revision procedure can be faced with a difficult decision. If a new ceramic ball is placed on a damaged metal taper, there is a much greater chance of fracture of the new ball.[26] On the other hand, if a metal ball is used, fragments of ceramic left in the tissues may subsequently become trapped between the ball and cup, triggering rapid wear of the metal ball and massive metallosis.[26,27] The safest option may be to replace the entire femoral component, but this can be problematic if the stem is well fixed.

Improvements in the quality of the alumina ceramic, particularly through hot isostatic pressing, have led to a reduction in fracture rate, to as low as one or two in 10,000 patients.[15,28] This risk must be weighed against the potential benefits of very low wear and high biocompatability of alumina-alumina bearings. An alumina-alumina total hip replacement of a contemporary design recently received FDA approval.

Ceramic-Ceramic Bearings Involving Zirconia

Due to its finer grain structure, yttria-stabilized zirconia ceramic is about 73% stronger than alumina, and therefore provides a greater margin of safety against fracture, particularly with smaller diameter femoral heads.[29-31] However, at high temperatures in a wet environment, zirconia can undergo a phase transformation that substantially weakens the material, roughens the surface, and degrades its wear properties.[18,32,33] While zirconia is highly stable at physiologic temperatures,[31-34] zirconia components should not be steam-autoclaved.

Zirconia has been used primarily as a femoral ball articulating against a polyethylene acetabular cup. The advisability of using zirconia balls against cups of alumina or zirconia is still a subject of some controversy.[35,36] Most recently, ceramics have been fabricated using a mixture of zirconia and alumina in various ratios. The resultant mixed-oxides ceramics appear to combine the high strength of zirconia with the thermal stability of alumina, and preliminary tests on a pin-on-disk machine and a hip joint simulator have indicated excellent wear resistance.[37,38]

New Forms of Polyethylene

Direct Compression Molded Components

In retrospective studies of shelf-stored and retrieved implants, components that were fabricated from block-molded or net-shape–molded polyethylene and gamma-sterilized and stored in air have undergone substantially less oxidation than similarly treated extruded-machined components.[39,40] Consequently, several suppliers have reintroduced net-shape molding of components. Specialized molding techniques can include hot isostatic pressing of the polyethylene powder in an argon atmosphere, and special molding protocols that can provide polyethylene with a specified crystallinity and stiffness.[41,42] However, because manufacturers no longer irradiate and store the polyethylene in air, the marked differences in oxidation levels between molded and extruded-machined components may no longer exist with current techniques of manufacturing and use.

Elevated Cross-linked and Thermally Stabilized Polyethylenes

Over the past few years, a number of laboratory wear simulations have demonstrated that the wear rate of polyethylene cups decreases markedly with an increasing level of radiation-induced cross-linking.[43-45] Although the baseline wear rate differed among various simulators due to systematic differences in the load, sliding distance per cycle, and other factors, the dose-wear curve is remarkably consistent among different laboratories and with varying cross-linking techniques (Fig. 2). The greatest reduction per Mrad occurs as the dose is increased from 0 to about 5 Mrads, with progressively less improvement at higher doses and no additional benefit above 15 to 20 Mrads (Fig. 2). While this dose-wear relationship was the basis for the recent development of a variety of intentionally cross-linked polyethylenes, the developers have arrived at very different conclusions as to the appropriate dose and other processing parameters for optimizing the clinical performance of a polyethylene implant.

Marathon™ and XLPE™ Gamma Radiation Cross-linked and Remelted Polyethylenes

In the Marathon™ process,[44] extruded bars of polyethylene are cross-linked by exposing them to 5 Mrads of gamma radiation, about 20% above the maximum dose used for conventional gamma sterilization. The bars are then heated to 155° C for 24 hours, followed by slow cooling to room temperature. When polyethylene is heated above the melt temperature, it is transformed from a partially crystalline solid to a totally amorphous solid. Because the uncombined free radicals generated during the gamma irradiation are trapped primarily in the crystalline regions, heating above the melt temperature frees them to combine with each other, forming additional

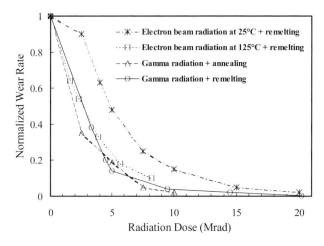

Figure 2 *Reduction in wear with increased level of cross-linking. The curves for the two gamma-radiation cross-linked polyethylenes were produced on hip joint simulators in two different laboratories.[43,44] The curves for the two electron-beam cross-linked polyethylenes were produced on a bidirectional pin-on-disk machine in a third laboratory.[45] Since two wear machines may produce different wear magnitudes for the same material due to differences in the applied load and the sliding distance per cycle, the original data were normalized by dividing by the wear rate for 0 Mrads (no cross-linking) obtained in each test. (Reproduced with permission from McKellop HA: Bearing surfaces in total hip replacements: State of the art and future developments. Instr Course Lect 2001;50:165-179.)*

crosslinks and—more importantly—minimizing the potential for long-term oxidative degradation. The acetabular cup is then machined from the central portion of the cross-linked–remelted bar, thereby removing the surface-oxidized material, and is sterilized using gas plasma rather than gamma radiation to avoid increasing the level of cross-linking or reintroducing residual free radicals. Under ideal clean test conditions, wear is reduced by about 85% (Fig. 2). Marathon™ has also demonstrated less wear than conventional polyethylene against severely roughened femoral balls,[46] as have Durasul™, Longevity™, and Crossfire™. XLPE™ cross-linked polyethylene is fabricated in much the same manner as Marathon™, except that the cross-linking dose is 10 Mrads and the final sterilization is done with ethylene oxide.[47]

Longevity™ Electron Beam Cross-linked and Remelted Polyethylene
In the Longevity™ process, compression-molded blocks of polyethylene are cross-linked by exposure to 10 Mrads of a 10-MeV electron beam,[46] which drives the cross-linking energy into the polyethylene about 2500 times faster than gamma radiation (ie, in seconds rather than hours). After cross-linking, the polyethylene is heated above the melt temperature for about 2 hours to extinguish the free radicals, and then machined into cups and sterilized with gas plasma.

Durasul™ Electron Beam Cross-linked and Remelted Polyethylene

In the Durasul™ process, the polyethylene is machined from compression-molded blocks into short segments or "pucks," which are cross-linked from both sides (to improve the uniformity of the cross-linking) with a 10-MeV electron beam to a total of 9.5 Mrads. In addition, the pucks are preheated to about 125° C while being electron beam cross-linked; this provides greater wear resistance (Fig. 2) and less reduction in elongation to break and toughness than electron beam cross-linking at room temperature.[49] After cross-linking, the Durasul™ cups are remelted to remove free radicals and sterilized with ethylene oxide.[50]

Crossfire™ and Aeonian™ Gamma Cross-linked and Annealed Polyethylenes

In the Crossfire™ process,[51] extruded bars of polyethylene are cross-linked by exposure to 7.5 Mrads of gamma radiation. The bars are then annealed (rather than melted) by heating them to just below the melt temperature for a proprietary duration. Cups are then machined from the bars, sealed in nitrogen, and sterilized by exposure to about 3 Mrads of gamma radiation (ie, to a total gamma dose of about 10.5 Mrads). Because no thermal treatment is applied to extinguish free radicals after the final gamma sterilization, Crossfire™ has more residual free radicals than implants sterilized with 2.5 to 4 Mrads of gamma radiation in the past. However, as with other currently gamma-sterilized polyethylenes, the manufacturer recommends storing Crossfire™ in their nitrogen-filled packages, and for only a limited time before implantation, in order to minimize oxidative degradation. The processing of Aenoian™ cross-linked polyethylene parallels that of Crossfire™ except that the initial cross-linking dose is about 3.5 Mrads, giving a total of 6 to 7.5 Mrads after final sterilization in nitrogen.

Relevance

Among those who advocate elevated cross-linking, the dose used represents that manufacturer's approach to balancing reduced wear against the need to maintain other mechanical properties, such as fracture toughness, above that needed for acceptable clinical performance. Those using the high levels of cross-linking (9.5 to 11 Mrads), about 2.5 to 3 times the maximum dose used historically to sterilize polyethylene components, maintain that the additional 5% to 10% improvement in wear over that provided by a moderate dose (Fig. 2), justifies the corresponding reduction in other physical properties. In contrast, advocates of a moderate cross-linking dose (such as 5 Mrads) maintain that the corresponding reduction in wear (about 85% compared to conventional polyethylene [Fig. 2]), if realized clinically, will be sufficient while avoiding the unnecessary reduction of other physical properties.

Because the polyethylene components fabricated by the historical process of gamma-sterilization in air are no longer marketed, surgeons must choose from among the new polyethylenes (no cross-linking, cross-linked as a byproduct of gamma sterilization, or intentionally cross-linked at moderate to high levels) or modern metal-metal or ceramic-ceramic bearings. The

Table 1 Advantages and disadvantages of current bearing choices

Bearing Combination	Advantages	Disadvantages
Alumina on alumina	Usually very low wear High biocompatability	Sometimes high wear Component fracture Higher cost Technique-sensitive surgery
Cobalt-chrome on cobalt-chrome	Usually low wear Can self-polish moderate surface scratches	Question of long-term local and systemic reactions to metal debris and/or ions
Hardened cobalt-chrome on polyethylene	Some additional protection against third-body abrasion	Hardened layer can wear off Higher cost
Ceramics on polyethylene	Lower wear of polyethylene than with conventional metal-polyethylene Some additional protection against third-body abrasion	Component fracture Difficulty of revision (eg, if Morse taper is damaged) Higher cost
Polyethylene sterilized with ethylene oxide or gas plasma	No short- or long-term oxidation	No cross-linking so polyethylene wear not minimized
Polyethylene sterilized with gamma in low oxygen	Some cross-linking, some wear reduction	Polyethylene wear not minimized Residual free radicals (long-term oxidation?)
Cross-linked, thermally stabilized polyethylene	Minimal polyethylene wear rate No short- or long-term oxidative degradation	Newest of low-wear bearing combinations, only early clinical results available Questions remain as to optimum cross-linking level and optimum method for thermal stabilization (melting versus annealing)

Adapted with permission from McKellop HA: Bearing surfaces in total hip replacements: State of the art and future developments. Instr Course Lect 2001;50:165-179.

advantages and disadvantages of each choice are summarized in Table 1. The surgeon must make an individual assessment of the risk-benefit ratio of the new bearing combinations.

Future Directions for Research

For metal-on-metal bearings, research may be directed toward appropriate modification of bearing surfaces to minimize run-in wear and micropitting, possibly through the use of high-tolerance, low-clearance bearings that achieve fluid-film separation of the surfaces. Identification of any adverse local or systemic effects of metal ion release is also an important research goal. Research in ceramic-ceramic bearings could target increasing the strength and toughness to further reduce the incidence of fracture, particularly in cases of damage due to dislocation/subluxation. Reductions in wear of metal-polyethylene bearings might be obtained through the use of special surface treatments or coatings on the femoral ball, such as ion-implanting or diamond-like coatings, although the cost-benefit ratio of these treatments may be an issue if the very low wear of the cross-linked polyethylenes against conventional metal heads demonstrated in the laboratory testing is realized in clinical use. For polyethylene, future research may focus on identifying the optimum level of cross-linking and the optimum process for thermal stabilization to minimize wear and oxidation while preserving fully adequate mechanical properties over decades of use in vivo. For each of these materials, close monitoring of clinical performance and accurate quantification of wear rates is essential for early recognition of unforeseen problems.

References

1. Schmalzried TP, Dorey FJ, McKellop H: The multifactorial nature of polyethylene wear in vivo. *J Bone Joint Surg Am* 1998;80:1234-1242.
2. Sauer WL, Anthony ME: Predicting the clinical wear performance of orthopaedic bearing surfaces, in Jacobs JJ, Craig TL (eds): *Alternative Bearing Surfaces in Total Joint Replacement.* West Conshohocken, PA, American Society for Testing and Materials, 1998, pp 1-29.
3. Wagner M, Wagner H: Preliminary results of uncemented metal on metal stemmed and resurfacing hip replacement arthroplasty. *Clin Orthop* 1996;329S:S78-S88.
4. Schmalzried TP, Fowble VA, Ure KJ, Amstutz HC: Metal on metal surface replacement of the hip: Technique, fixation, and early results. *Clin Orthop* 1996;329S:S106-S114.
5. McMinn D, Treacy R, Lin K, Pynsent P: Metal on metal surface replacement of the hip: Experience of the McMinn prothesis. *Clin Orthop* 1996;329S:S89-S98.
6. Schmidt M, Weber H, Schon R: Cobalt chromium molybdenum metal combination for modular hip prostheses. *Clin Orthop* 1996;329S:S35-S47.
7. Weber BG: Experience with the Metasul total hip bearing system. *Clin Orthop* 1996;329S:S69-S77.
8. Streicher RM, Semlitsch M, Schon R, Weber H, Rieker C: Metal-on-metal articulation for artificial hip joints: Laboratory study and clinical results. *J Eng Med* 1996;210:223-232.
9. Chan FW, Bobyn JD, Medley JB, Krygier JJ, Tanzer M: The Otto Aufranc Award: Wear and lubrication of metal-on-metal hip implants. *Clin Orthop* 1999;369:10-24.
10. Rieker CB, Weber H, Schon R, Windler M, Wyss UP: Development of the metasul articulations, in Rieker MW, Wyss C (eds): *Metasul: A Metal-on-Metal Bearing.* Bern, Switzerland, Hans Huber, 1999, pp 15-21.
11. Rieker CB, Köttig P, Schön R, Windler M, Wyss UP: Clinical tribological performance of 144 metal-on-metal hip articulations, in Rieker W, Wyss U (eds): *Metasul: A Metal-on-Metal Bearing.* Bern, Switzerland, Hans Huber, 1999, pp 83-91.
12. Campbell P, McKellop H, Alim R, et al: Metal-on-metal hip replacements: Wear performance and cellular response to wear particles, in Disegi JA, Kennedy RL, Pilliar R (eds): *Cobalt-Based Alloys for Biomedical Applications.* West Conshohocken, PA, American Society for Testing and Materials, 1998.
13. Park S-H, McKellop H, Lu B, Chan F, Chiesa R: Wear morphology of metal-metal implants: Hip simulator tests compared with clinical retrievals, in Jacobs JJ Craig TL (eds): *Alternative Bearing Surfaces in Total Joint Replacement.* West Conshohocken, PA, American Society for Testing and Materials, 1998, pp 129-143.

14. Dorr LD, Wan Z, Longjohn DB, Dubois B, Murken R: Total hip arthroplasty with use of the Metasul metal-on-metal articulation: Four to seven-year results. *J Bone Joint Surg Am* 2000;82:789-798.

15. Clarke I, Willmann G: Structural ceramics in orthopaedics, in Cameron H (ed): *Bone Implant Interface*. St Louis, MO, Mosby, 1994, pp 203-252.

16. Willmann G: New generation ceramics, in Willmann G, Zweymuller K (eds): *Bioceramics in Hip Joint Replacement*. Proceedings 5th International Ceram Tec Symposium. Stuttgart, Germany, Georg Thieme, 2000, pp 127-135.

17. Devane PA, Horne JG: Assessment of polyethylene wear in total hip replacement. *Clin Orthop* 1999;369:59-72.

18. Allain J, Le Mouel S, Goutallier D, Voison McAllain J: Poor eight-year survival of cemented zirconia-polyethylene total hip replacements. *J Bone Joint Surg Br* 1999;81:835-842.

19. Mittelmeier H, Heisel J: Sixteen-years' experience with ceramic hip prostheses. *Clin Orthop* 1992;282:64-72.

20. Nizard RS, Sedel L, Christel P, et al: Ten-year survivorship of cemented ceramic-ceramic total hip prosthesis. *Clin Orthop* 1992;282:53-63.

21. Garcia-Cimbrelo E, Martinez-Sayanes JM, Minuesa A, Munuera L: Mittelmeier ceramic-ceramic prosthesis after 10 years. *J Arthroplasty* 1996;11:773-781.

22. Richter HG, Willmann G, Weick K: Improving the reliability of the ceramic-on-ceramic wear couple in THR, in Jacobs JJ, Craig TL (eds): *Alternative Bearing Surfaces in Total Joint Replacement*. West Conshohocken, PA, American Society for Testing and Materials, 1998, pp 173-185.

23. Nevelos JE, Fisher J, Ingham E, Doyle C, Nevelos AB: Examination of alumina ceramic components from Mittelmeier total hip arthroplasties. *Trans Orthop Res Soc* 1998:23:219.

24. Bergman NR, Young DA: The rationale, short-term outcome and early complications of a ceramic couple in total hip arthroplasty, in Sedel L, Willmann G (eds): *Reliability and Long-Term Results of Ceramics in Orthopaedics*. Stuttgart, Germany, Georg Thieme Verlag, 1999, pp 52-56.

25. Yoon TR, Rowe SM, Jung ST, Seon KJ, Maloney WJ: Osteolysis in association with a total hip arthroplasty with ceramic bearing surfaces. *J Bone Joint Surg Am* 1998;80:1459-1468.

26. Allain J, Goutallier D, Voisin MC, Lemouel S: Failure of a stainless-steel femoral head of a revision total hip arthroplasty performed after a fracture of a ceramic femoral head. *J Bone Joint Surg Am* 1998;80:1355-1360.

27. Kempf I, Semlitsch M: Massive wear of a steel ball head by ceramic fragments in the polyethylene acetabular cup after revision of a total hip prosthesis with fractured ceramic ball. *Arch Orthop Trauma Surg* 1990;109:284-287.

28. Fritsch EW, Gleitz M: Ceramic femoral head fractures in total hip arthroplasty. *Clin Orthop* 1996;328:129-136.

29. Drouin JM, Cales B, Chevalier J, Fantozzi G: Fatigue behavior of zirconia hip joint heads: Experimental results and finite element analysis. *J Biomed Mater Res* 1997;34:149-155.

30. Cales B, Stefani Y: Yttria-stabilized zirconia for improved orthopaedic prostheses, in Wise DL, Trantolo DJ, Yaszemski MJ, Gresser JD, Schwartz ER (eds): *Encyclopedic Handbook of Biomaterials and Bioengineering*. New York, NY, Marcel Dekker, 1995.

31. Piconi C, Maccauro G: Zirconia as a ceramic biomaterial. *Biomaterials* 1999;20:1-25.

32. Birkby I, Harrison P, Stevens R: The effect of surface transformation on the wear behavior of zirconia TZP ceramics. *J Euro Ceramics Soc* 1989;5:37-45.

33. Piconi C, Burger W, Richter HG, et al: Y-TZP ceramics for artificial joint replacements. *Biomaterials* 1998;19:1489-1494.

34. Cales B, Stefani Y, Lilley E: Long-term in vivo and in vitro aging of a zirconia ceramic used in orthopaedy. *J Biomed Mater Res* 1994;28:619-624.

35. Fruh HJ, Willmann G, Pfaff HG: Wear characteristics of ceramic-on-ceramic for hip endoprostheses. *Biomaterials* 1997;18:873-876.

36. Cales B, Chevalier J: Wear behavior of ceramic pairs compared on different testing configurations, in Jacobs JJ, Craig TL (eds): *Alternative Bearing Surfaces in Total Joint Replacement.* West Conshohocken, PA, American Society for Testing and Materials, 1998, pp 186-196.

37. Affatato S, Testoni M, Cacciari GL, Toni A: Mixed oxides prosthetic ceramic ball heads: Part 1. Effect of the ZrO2 fraction on the wear of ceramic on polythylene joints. *Biomaterials* 1999;20:971-975.

38. Affatato S, Testoni M, Cacciari GL, Toni A: Mixed-oxides prosthetic ceramic ball heads: Part 2. Effect of the ZrO2 fraction on the wear of ceramic on ceramic joints. *Biomaterials* 1999;20:1925-1929.

39. Currier BH, Currier JH, Collier JP, Mayor MB: Effect of fabrication method and resin type on performance of tibial bearings. *J Biomed Mater Res* 2000;53:143-151.

40. Gillis A, Furman B, Li S: Influence of ultra high molecular weight polyethylene resin type and manufacturing method on real time oxidation. *Trans Orthop Res Soc* 1998;23:360.

41. Kurtz SM, Muratoglu OK, Evans M, Edidin AA: Advances in the processing, sterilization, and crosslinking of ultra-high molecular weight polyethylene for total joint arthroplasty. *Biomaterials* 1999;20:1659-1688.

42. Walsh H, Gillis A, Furman B, Li S: Factors that determine the oxidation resistance of molded 1900: Is it the resin or the molding? *Trans Orthop Res Soc* 2000;25:543.

43. Wang A, Essner A, Polineni VK, Stark C, Dumbleton JH: Lubrication and wear of ultra-high molecular weight polyethylene in total joint replacements. *Tribology Int* 1998;31:17-33.

44. McKellop HA, Shen FW, Lu B, Campbell P, Salovey R: Development of an extremely wear resistant UHMW polyethylene for total hip replacements. *J Orthop Res* 1999;17:157-167.

45. Muratoglu OK, Bragdon CR, O'Connor DO, et al: Unified wear model for highly cross-linked ultra-high molecular weight polyethylenes (UHMWPE). *Biomaterials* 1999;20:1463-1470.

46. McKellop H, Shen F-W, DiMaio W, Lancaster J: Wear of gamma-cross-linked polyethylene acetabular cups against roughened femoral balls. *Clin Orthop* 1999;369:73-82.

47. Greenwald AS, Ries MD, Duncan CP, Jacobs JJ, Stulberg BN: New polys for old: Contribution or caveat? Scientific Exhibit 057 presented at the 67th Annual Meeting of the American Academy of Orthopaedic Surgeons, March 2000.

48. Laurent M, Yao JQ, Bhambri SK, et al: High cycle wear of highly cross-linked UHMWPE acetabular liners evaluated in a hip simulator. *Trans Orthop Res Soc* 2000;25:567.

49. Muratoglu OK, Bragdon CR, O'Connor DO, et al: The effect of temperature and radiation cross-linking on UHMWPE for use in total hip arthroplasty. *Trans Orthop Res Soc* 2000:25:547.

50. Muratoglu OK, Harris WH, Delaney H, et al: The development of an in vitro hip simulator model for fatigue failure: Application to conventional and highly cross-linked UHMWPE. *Trans Orthop Res Soc* 2000:25:548.

51. Manley MT, Capello WN, D'Antonio JA, Taylor SK, Wang A: Reduction in wear in total hip replacement: Highly crosslinked polyethylene acetabular liners versus ceramic/ceramic bearings. Scientific Exhibit 022 presented at the 67th Annual Meeting of the American Academy of Orthopaedic Surgeons, March 2000.

Index

A

Abrasive-adhesive wear
  in hip components, 139
  work to failure and, 143
Abrasive wear, 177-178, 180
Accelerated fatigue wear, 179
Accelerated wear, prompt and accurate detection
    of, 28
Acetabular component
  alignment of, to polyethylene wear, 20
  correlation between activity and wear, 14
  femoral head penetration into, 25–26
Acetabular cups
  fracture of, 179, 198–199
  impingement and anteversion of, 20
  machining, 210
  particle generation in, 141
Acetabular osteolysis, 44–45, 50–51
Acetabular screws, removal of, in revision
    surgery, 44
Acetabular shell removal, 44
Acetabulum, cement fixation in the, 4
Acetylcysteine, effect of, on
    transcription factors, 74
Activity, correlation between acetabular compo-
    nent wear and, 14
Adhesive wear, 177-178, 180
Aeonian™ gamma cross-linked and annealed
    polyethylenes, 211
Age
  effect of corrosion, on cellular response, 101
  implant wear and, 13-15
Aggressive granulomatosis, 61
Alendronate
  in preventing osteolysis, 46
  in reducing bone loss, 132
Allografts
  bulk, 56
  cancellous
    in restoring bone stock, 56
    in treating acetabular osteolysis, 44–45
  circumferential, 55–56
  in treating bone loss, 132
Alternative bearing materials, evidence for using,
    206–213
Alumina-alumina bearings, 208
Alumina/zirconia ceramic composites, 203
Aluminum-associated osteomalacia, 125
Ankle joints, need for research on, 202
Antibiotics, incorporation of, in bone cement,
    131
Archard relationship, 171
Archard-type computational model, parametric
    trials using, 172
Articular side wear
  need for retrieval research in, 11
  as source of implant debris, 9
Ascorbate, effect of, on transcription factors, 74
Aseptic loosening, 4

as cause of total hip replacement failure, 61
characteristics of, 62
cyclic loading as cause of, 69
cytofluorographic studies in investing, 62
interface tissue in, 61–62
macrophage-osteoclast differentiation in, 77
new technologies in understanding
    pathogenesis of, 119
periprosthetic bone loss as cause of, 38, 69
pseudosynovial fluid in, 62
as reason for revision arthroplasty, 8
wear particles in, 91
Aspirin, effect of, on transcription factors, 74
Autopsy retrievals following revision, need for,
    57

B

Backside wear
  knee simulator studies in examining, 11, 165,
    171, 201
  need for retrieval research in, 11
Bacteria in loose total hip replacements, 61
Barium sulfate, use of, as contrast media, 62
Bearings. *See also* Ceramic-on-ceramic bearings;
Ceramic-on-polyethylene bearings; Metal-on
metal bearings; Metal-on-polyethylene bearings;
Mobile bearings
  evidence for using alternative materials for,
    206–213
  surface design in determining sliding distance,
    156
Belt finishing, 160
Biologic factors, xvi
  experimental approaches in investigating effects
    of particles, 114–119
  features of wear particles in determining
    adverse reactions, 94–102
  host factors in determining responses to wear
    particles, 86–91
  identifying biologic markers of wear, 124–128
  local and systemic reactions to wear debris,
    61–69
  mediators of, 71–79
  role of endotoxin and fluid pressure
    in osteolysis, 106–111
  in treating osteolysis, 130–133
Biologic markers of wear, identifying, 124–128
Biologic reactions, features of wear particles in
    determining adverse, 94–102
  form, 101
  future research, 102
  material, 99–101
  particle concentration, 98–99
  particle size, 94-98
  relevance, 101–102
Biomechanical loading, 188
Bisphosphonates
  need for research on, 69
  in preventing osteolysis, 46